Nonlinear Optics

WILEY SERIES IN MICROWAVE AND OPTICAL ENGINEERING

KAI CHANG, Editor
Texas A & M University

Nonlinear Optics

E. G. SAUTER

Institute for High Frequency Technique and Quantum Electronics
University of Karlsruhe

A WILEY-INTERSCIENCE PUBLICATION

JOHN WILEY & SONS, INC.

NEW YORK / CHICHESTER / BRISBANE / TORONTO / SINGAPORE

This text is printed on acid-free paper.

Copyright © 1996 by John Wiley & Sons, Inc.

All rights reserved. Published simultaneously in Canada.

Reproduction or translation of any part of this work beyond
that permitted by Section 107 or 108 of the 1976 United
States Copyright Act without the permission of the copyright
owner is unlawful. Requests for permission or further
information should be addressed to the Permissions Department,
John Wiley & Sons, Inc., 605 Third Avenue, New York, NY
10158-0012

Library of Congress Cataloging in Publication Data is available.
0-471-14860-1
Printed in the United States of America
10 9 8 7 6 5 4 3 2 1

In memory of my parents

Contents

Preface

This textbook originated in a course of lectures in nonlinear optics that I have given for several years now at the University of Karlsruhe. Many years ago, my scientific career started with a doctoral thesis on nonlinear vibrations. I next turned my activities to high-energy physics. But now, I return to my first love: nonlinearities in optics.

The book provides an introduction to the phenomena of nonlinear optics in classical terms, avoiding quantum theory to make the book accessible to a readership not having a quantum theoretical background. This restriction forced me to discuss in some places classical phenomenological models and forbids, for example, a detailed treatment of spontaneous processes.

The background necessary for understanding this text is provided in most basic optics textbooks, some of which are listed below. The book is intended as a textbook for students in engineering and physics, as well as a reference for the practitioner who must use nonlinear optics as a new tool in his or her work. The text is written in a reasonably self-contained manner and will, hopefully, provide the student with the knowledge necessary to understand research articles on the topic.

The material can be covered in a one-semester course. The first chapter provides a general introduction to the basic relations between polarization and electric field strength. Symmetry relations help to reduce the number of independent components of the susceptibility tensors χ. A survey of the different nonlinear effects of polarizations of order two and three follows. Since we do not use quantum theory, we need a classical model that allows us to make statements on the structure of χ. This is done in the next section. The chapter ends with a discussion of the relation between the real and imaginary parts of the susceptibility.

In Chapter 2, we treat the wave propagation of complex phasors in nonlinear media and approximate the differential equations by assuming a slowly varying amplitude (or envelope). The wave propagation in anisotropic media is also treated. Conservation of energy and momentum in nonlinear processes is contained in the Manley–Rowe equations and phase matching, respectively.

In Chapter 3, the discussion of individual nonlinear effects starts with the linear electro-optic or Pockels effect (Pockels, 1893). The Faraday effect (Faraday, 1845) is an example in which polarization depends on a magnetic field. In Chapter 4, we treat second harmonic generation (Franken et al., 1961). Parametric effects are discussed in Chapter 5: sum and difference frequency mixing, frequency up and down conversion, parametric amplification, and oscillation. We begin the discussion of four-wave mixing phenomena with the Raman effect (spontaneous Raman effect: Raman, 1928; induced Raman effect: Woodbury and Ng, 1962) and the Brillouin effect (spontaneous Brillouin effect: Brillouin, 1922; stimulated Brillouin effect: Chiao et al., 1964) in Chapter 6 and continue in Chapter 7 with the quadratic electro-optic or optical Kerr effect (Kerr, 1875). As important applications, we discuss here bistability and self-focusing. In Chapter 8, the general theory of four-wave mixing is treated with phase conjugation as an application.

In Chapter 9, we investigate the propagation of light pulses first in linear media with an emphasis on pulse broadening and compression. After this, propagation in nonlinear media follows with self-phase modulation and modulation instability as applications. The important special case of solitons is discussed in Chapter 10, an old subject in principle, since J. Scott Russel first observed solitons as water waves in 1834 (described in 1844), but only in the modern laser era was attention again focused on this topic. In Chapter 11, we discuss both the harmful and useful nonlinear effects in glass fibers. While the discussions in all the other chapters will remain essentially unchanged, although some new applications may be found in the future, Chapter 11 will need revision owing to the rapid developments in the field of optical communications.

The appendices contain sections on notation and different systems of units, a table of susceptibility tensors, proof of an overall permutation symmetry, and a short description of the solution of nonlinear differential equations by inverse scattering theory.

Each chapter concludes with a collection of problems that deal with supplements to the text, analytical derivations, and applications. The text also includes numerous examples, some of which have been worked out. The reading lists at the end of each chapter contain textbooks and monographs, only in rare cases supplemented by special research articles.

Besides the omission of quantum mechanical effects, the expert will miss a discussion of spontaneous light scattering, Rayleigh scattering, photorefractive effects, nonlinear effects in semiconductors, and so on, the treatment of which would have enlarged the book too much. Some effects, as for example, nonlinear spectroscopy, are only touched upon so that the student will get at least an idea of where these effects would have their place.

In contrast to many research papers, I use here the international SI system of units. This should be convenient for students of applied physics and engineering programs. However, Appendix B gives sufficient information on how to convert these units to cgs. We choose a time dependence of the form $\exp(j\omega t)$ for stationary processes. People who prefer the negative (but you could also use "other sign") sign in the exponent, should take the complex conjugate of all functions.

In view of the great number of contributors in this area I must apologize for not being able to cite all of them in this book. I owe a great debt of thanks to Prof. G. K. Grau of the University of Karlsruhe, from whom I learned the modern aspects of nonlinear optics. His book on Quantum Electronics which unfortunately appears only in a German edition proved a steady source of knowledge for me. Without the encouragement of my friend, Prof. S. R. Seshadri of the University of Wisconsin, Madison, I would not have undertaken the task of writing this book. I am most grateful to him. Thanks are also due to my students who have helped to eliminate some of the obscurities and errors in the class notes on which this book is based. I am indebted to Mrs. D. Goldmann for her help in preparing the manuscript and to Mrs. I. Kober for drawing all the figures very carefully.

E. G. SAUTER

Karlsruhe, Germany

READING LIST:

M. Born and E. Wolf. *Principles of Optics*. Pergamon Press, Oxford, 1964.

R. W. Ditchburn, *Light*, 3rd ed. Academic Press, London, 1976.

G. R. Fowles. *Introduction to Modern Optics*. Holt, Rinehart and Winston, New York, 1968.

A. K. Ghatak. *An Introduction to Modern Optics*. Tata McGraw-Hill, Bombay, 1971.

E. Hecht and A. Zajac. *Optics*, 2nd ed. Addison-Wesley, Reading, MA, 1987.

F. A. Jenkins and H. E. White. *Fundamentals of Optics*, 4th ed. McGraw-Hill, New York, 1976.

M. V. Klein and T. E. Furtak. *Optics*, 2nd ed. John Wiley & Sons, New York, 1986.

S. G. Lipson and H. Lipson. *Optical Physics*, 2nd ed. Cambridge University Press, London, 1981.

R. S. Longhurst. *Geometrical and Physical Optics*, 3rd ed. Longman, London, 1973.

J. R. Mayer-Arendt. *Introduction to Classical and Modern Optics*, 3rd ed. Prentice-Hall, Englewood Cliffs, NJ, 1989.

A. Nussbaum and R. A. Phillips. *Contemporary Optics for Scientists and Engineers*. Prentice-Hall, Englewood Cliffs, NJ, 1976.

F. L. Pedrotti and L. S. Pedrotti. *Introduction to Optics*. Prentice-Hall, Englewood Cliffs, NJ, 1987.

Nonlinear Optics

CHAPTER ONE

Electric Field and Polarization

This and the following chapter constitute the basis for the treatment of special nonlinear effects in the stationary case in Chapters 3 through 8. We discuss quite generally the relation between the electric field and polarization and we define susceptibility tensors. Symmetry properties allow one to reduce the number of independent elements of the susceptibility tensors. The different specific non-linear effects are derived up to polarization of the third order. Since we avoid quantum theory in our treatment, we have to model the nonlinear medium by a classical anharmonic oscillator to obtain the structure of the susceptibility tensors. In the linear case, the real and imaginary parts of the susceptibilities are connected by a Hilbert transform (Section 1.5), but in the nonlinear case, these dispersion relations are not very valuable. The notation is clarified in Appendix A.

1.1 THE RELATION BETWEEN E AND P; SUSCEPTIBILITY

The constitutive equation between electric displacement $\mathbf{D}$ and electric field $\mathbf{E}$ can be written in two different forms:

$$\mathbf{D} = \varepsilon\mathbf{E} = \varepsilon_0\mathbf{E} + \mathbf{P}_L \qquad (1.1)$$

In the first form, $\mathbf{D}$ is traced back to the electric field $\mathbf{E}$ using the permittivity $\varepsilon = \varepsilon_0\varepsilon_r$; in the second form, $\mathbf{D}$ consists of a part $\varepsilon_0\mathbf{E}$ from vacuum and a part $\mathbf{P}_L$ from matter (linear polarization). In anisotropic media, $\mathbf{D}$ and $\mathbf{E}$ are not parallel in general; ε is then a tensor of rank 2: $\varepsilon \rightarrow \boldsymbol{\varepsilon}$. Equation (1.1) solved for $\mathbf{P}_L$ leads to the definition of a first-order susceptibility tensor $\boldsymbol{\chi}^{(1)}$ (tensor of rank 2):

$$P_{Li} = \sum_{j=1}^{3}(\varepsilon_{ij} - \varepsilon_0\delta_{ij})E_j =: \varepsilon_0\sum_j \chi_{ij}^{(1)}E_j \qquad (1.2)$$

1

Here, the polarization $\mathbf{P}_L$ depends linearly on the field:

$$\mathbf{P}_L(\mathbf{E}) =: \mathbf{P}^{(1)}(\mathbf{E})$$

For strong fields, the preceding linear relation breaks down; the *most general relation* between $\mathbf{P}$ and $\mathbf{E}$ contains besides linear terms also higher powers of $\mathbf{E}$. This leads to a nonlinear contribution to polarization:

$$\mathbf{D} = \varepsilon\mathbf{E} + \mathbf{P}_{NL} = \varepsilon_0\mathbf{E} + \mathbf{P}_L + \mathbf{P}_{NL} = \varepsilon_0\mathbf{E} + \mathbf{P} = \varepsilon(\mathbf{E}^0, \mathbf{E}^1, \mathbf{E}^2, \ldots)\cdot\mathbf{E} \quad (1.3)$$

or

$$\mathbf{P} = \mathbf{P}_L + \mathbf{P}_{NL} = \mathbf{P}^{(1)}(\mathbf{E}) + \mathbf{P}^{(2)}(\mathbf{E}^2) + \mathbf{P}^{(3)}(\mathbf{E}^3) + \ldots \quad (1.4)$$

Explicitly, the notation in Eq. (1.4) denotes the most general dependence of a time function $\mathbf{P}$ on another time function $\mathbf{E}$, for example,

$$\mathbf{P}^{(1)}(t) = \mathbf{P}^{(1)}(\mathbf{E}^1) = \int_{-\infty}^{+\infty} d\tau\, \mathbf{q}^{(1)}(t,\tau)\cdot\mathbf{E}(\tau) \quad (1.5)$$

$$\mathbf{P}^{(2)}(t) = \mathbf{P}^{(2)}(\mathbf{E}^2) = \int_{-\infty}^{+\infty} d\tau_1 \int_{-\infty}^{+\infty} d\tau_2\, \mathbf{q}^{(2)}(t,\tau_1,\tau_2) : \mathbf{E}(\tau_1)\mathbf{E}(\tau_2) \quad (1.6)$$

[for the notation, see Eq. (1.10) and Appendix A.]

Remarks:

1. The field $\mathbf{E}$ is here an arbitrary function of time. Later on, Eq. (1.40), we shall decompose it into a sum of p monochromatic fields $\mathbf{E}_r$ $(r = 1,\ldots,p)$ at different frequencies. The product $\mathbf{E}(\tau_1)\mathbf{E}(\tau_2)$, for example, in Eq. (1.6) presents itself then as a sum of terms representing different processes and containing additional prefactors $c^{(n)}$ (see Section 1.3).

2. The second-order polarization $\mathbf{P}^{(2)}$ describes a *three-wave mixing* [two E fields interact and generate a third (polarization) wave], the third-order polarization $\mathbf{P}^{(3)}$ similarly *four-wave mixing* (FWM). Higher nonlinearities are of no interest here.

3 As a matter of fact, one should also consider in the expansion of $\mathbf{P}$ in Eq. (1.4) terms with magnetic field, moreover the spatial and temporal derivatives of those fields. For example, one can describe the Faraday effect with a term proportional to $\mathbf{B}\cdot\mathbf{E}$, see Section 3.4.

$\mathbf{q}(t,\tau) = (q_{ij})$ from Eq. (1.5) is a tensor of rank 2, $\mathbf{q}^{(2)}(t,\tau_1,\tau_2) = (q_{ijk})$ from Eq. (1.6) a tensor of rank 3. By the independence of physical processes from the choice of an arbitrary origin of the time scale, the form of this tensors is determined. This *independence from time displacement* (time shift invariance) means an independence of the dynamic properties of the medium from the origin of time: If, for example, the polarization $P(t_2)$ is induced by the field $E(t_1)$, then the polarization $P(t_2 + T)$ is induced by $E(t_1 + T)$.

Example 1.1   Determine $\mathbf{q}(t, \tau)$, $\mathbf{q}^{(2)}(t, \tau_1, \tau_2)$.
At first, we replace t in Eq. (1.5) by $t + T$:

$$\mathbf{P}(t + T) = \int_{-\infty}^{+\infty} d\tau \, \mathbf{q}(t + T, \tau)\mathbf{E}(\tau) \tag{1.7}$$

On the other hand, if we shift $\mathbf{E}(\tau)$ in Eq. (1.5) by T to $\mathbf{E}(\tau + T)$, we obtain due to invariance with respect to time displacement $\mathbf{P}(t) \rightarrow \mathbf{P}(t + T)$:

$$\mathbf{P}(t + T) = \int_{-\infty}^{+\infty} d\tau \, \mathbf{q}(t, \tau)\mathbf{E}(\tau + T) = \int_{-\infty}^{+\infty} d\tau \, \mathbf{q}(t, \tau - T)\mathbf{E}(\tau) \tag{1.8}$$

A comparison with Eq. (1.7) yields

$$\mathbf{q}(t + T, \tau) = \mathbf{q}(t, \tau - T) \qquad \text{for all } t, T, \tau$$

We now set $t = 0$ and then $T = t$. This results in

$$\mathbf{q}(t, \tau) = \mathbf{q}(0, \tau - t)$$

that is, $\mathbf{q}$ depends only on the difference $\tau - t$. Therefore, we write

$$\mathbf{q}(t, \tau) =: \varepsilon_0 \chi^{(1)}(t - \tau) \tag{1.9}$$

Equation (1.6) is written in components:

$$P_i^{(2)}(t) = \sum_{j,k} \int_{-\infty}^{+\infty} d\tau_1 \int_{-\infty}^{+\infty} d\tau_2 \, q_{ijk}(t, \tau_1, \tau_2)E_j(\tau_1)E_k(\tau_2) \tag{1.10}$$

If we decompose q_{ijk} into a symmetric (S_{ijk}) and an antisymmetric (A_{ijk}) part with respect to a permutation of the pairs (j, τ_1) and (k, τ_2), we see that the antisymmetric part yields a vanishing contribution after multiplication with $E_j(\tau_1)E_k(\tau_2)$ in Eq. (1.10). So we can assume $A_{ijk} = 0$ without restricting generality (see Problem 1.2) and take q as symmetric in the pairs (j, τ_1), (k, τ_2):

$$q_{ijk}(t, \tau_1, \tau_2) = q_{ikj}(t, \tau_2, \tau_1) \tag{1.11}$$

Under time shift invariance, Eq. (1.10) transforms into

$$P_i^{(2)}(t+T) = \sum_{j,k} \int_{-\infty}^{+\infty} d\tau_1 \int_{-\infty}^{+\infty} d\tau_2 \; q_{ijk}(t,\tau_1,\tau_2)E_j(\tau_1+T)E_k(\tau_2+T)$$

$$= \sum_{j,k} \int_{-\infty}^{+\infty} d\tau_1 \int_{-\infty}^{+\infty} d\tau_2 \; q_{ijk}(t,\tau_1-T,\tau_2-T)E_j(\tau_1)E_k(\tau_2) \quad (1.12)$$

Comparison with Eq. (1.10), after we have replaced t there by $t+T$, yields

$$q_{ijk}(t+T,\tau_1,\tau_2) = q_{ijk}(t,\tau_1-T,\tau_2-T) \qquad \text{for all } t,T,\tau_1,\tau_2$$

We again set $t=0$ and then $T=t$. This yields

$$q_{ijk}^{(2)}(t,\tau_1,\tau_2) = q_{ijk}^{(2)}(0,\tau_1-t,\tau_2-t) =: \varepsilon_0 \chi_{ijk}^{(2)}(t-\tau_1,t-\tau_2) \qquad (1.13)$$

∎

With the expressions, Eqs. (1.9) and (1.13), for the **q**-tensors, Eqs. (1.5) and (1.6) transform into

$$P_i^{(1)}(\mathbf{x},t) = \varepsilon_0 \sum_j \int_{-\infty}^{+\infty} d\tau \; \chi_{ij}^{(1)}(t-\tau)E_j(\mathbf{x},\tau)$$

$$= \varepsilon_0 \sum_j \int_{-\infty}^{+\infty} d\tau \; \chi_{ij}^{(1)}(\tau)E_j(\mathbf{x},t-\tau) = \varepsilon_0 \sum_j \chi_{ij}^{(1)}(t) * E_j(\mathbf{x},t) \qquad (1.14)$$

($*$ means convolution!) and

$$P_i^{(2)}(\mathbf{x},t) = \varepsilon_0 \sum_{j,k} \int_{-\infty}^{+\infty} d\tau_1 \int_{-\infty}^{+\infty} d\tau_2 \; \chi_{ijk}^{(2)}(t-\tau_1,t-\tau_2)E_j(\mathbf{x},\tau_1)E_k(\mathbf{x},\tau_2)$$

$$= \varepsilon_0 \sum_{j,k} \int_{-\infty}^{+\infty} d\tau_1 \int_{-\infty}^{+\infty} d\tau_2 \; \chi_{ijk}^{(2)}(\tau_1,\tau_2)E_j(\mathbf{x},t-\tau_1)E_k(\mathbf{x},t-\tau_2) \qquad (1.15)$$

with the susceptibility tensors $\chi_{ij}^{(1)}(\tau)$ and $\chi_{ijk}^{(2)}(\tau_1,\tau_2)$ being real functions of time. Due to causality, we still have

$$\chi_{ij}^{(1)}(\tau) = 0 \qquad \text{for} \quad \tau < 0 \qquad (1.16)$$

and

$$\chi_{ijk}^{(2)}(\tau_1, \tau_2) = 0 \qquad \text{for } \tau_1 < 0 \text{ or } \tau_2 < 0 \qquad (1.17)$$

The structure of Eq. (1.14)—similarly that of Eq. (1.15)—can be understood by the following physical reasoning: The electric field (as cause, input) influences vibrations of the bound electrons in the atoms of the medium. This gives rise to a (time-delayed) polarization (as result, output). Or, the polarization at time t depends on all fields at earlier times via the response function $\chi(\tau)$.

The Fourier transform of Eq. (1.14) with respect to time yields immediately

$$P_i^{(1)}(\mathbf{x}, f) = \varepsilon_0 \sum_j \chi_{ij}^{(1)}(f)E_j(\mathbf{x}, f), \qquad \mathbf{P}_L(\mathbf{x}, f) = \varepsilon_0 \chi^{(1)}(f)\, \mathbf{E}(\mathbf{x}, f) \quad (1.18)$$

$$\chi_{ij}^{(1)}(f) = \int_{-\infty}^{\infty} \chi_{ij}^{(1)}(\tau)e^{-2\pi j f \tau}\, d\tau, \qquad \chi_{ij}^{(1)}(-f) = \chi_{ij}^{(1)*}(f) \qquad (1.19)$$

Thus, Eq. (1.2) holds true only in the frequency domain, not the time domain!

For a medium without memory, we set in Eq. (1.19)

$$\chi_{ij}^{(1)}(\tau) = \chi_{ij}^{(1)}\delta(\tau)$$

Therefore, $\chi_{ij}^{(1)}(f) = \chi_{ij}^{(1)}$, and is real and independent of frequency. In this case, we conclude from Eq. (1.18) that Eq. (1.2) also holds true in the time domain. But if $\chi_{ij}^{(1)}(f)$ is still almost constant (and therefore almost real), then Eq. (1.2) holds true at least approximately in the time domain. In general, we see that if

$$\chi_{ij}(f) = \chi'(f) + j\chi''(f)$$

is constant, independent of frequency, then it is real too, but from reality ($\chi'' = 0$) does not follow independence of frequency.

The definition of permittivity in the linear case takes place likewise in the frequency domain:

$$D_i(\mathbf{x}, f) = \sum_j \varepsilon_{ij}(f)E_j(\mathbf{x}, f) = \varepsilon_0 E_i(\mathbf{x}, f) + P_i^{(1)}(\mathbf{x}, f)$$

$$= \varepsilon_0 \sum_j [\delta_{ij} + \chi_{ij}^{(1)}(f)]E_j(\mathbf{x}, f) \qquad (1.20)$$

$$\varepsilon_{ij}(f) = \varepsilon_0[\delta_{ij} + \chi_{ij}^{(1)}(f)] \qquad (1.21)$$

$$\varepsilon(f) = \varepsilon_0[1 + \chi^{(1)}(f)] = \varepsilon_0 \varepsilon_r = \varepsilon_0 n^2 \text{ if isotropic}$$

Thus, Eq. (1.1) also holds true in the time domain only if $\varepsilon \cdot \mathbf{E}$ will be understood there as convolution! Since $\chi_{ij}^{(1)}(f)$ is complex in general, $\varepsilon_{ij}(f)$ has a non-vanishing imaginary part that is connected to the unavoidable losses in matter. Equation (1.21) also shows that a large linear susceptibility results in a large refractive index.

For a monochromatic signal of frequency f_1, we obtain from Eq. (1.18) for the complex phasors:

$$\widehat{P}_i^{(1)}(\mathbf{x},f_1) = \varepsilon_0 \sum_j \chi_{ij}^{(1)}(f_1)\widehat{E}_j(\mathbf{x},f_1) \tag{1.22}$$

The Fourier transform of Eq. (1.15) with respect to time for the quadratic nonlinearity is

$$P_i^{(2)}(\mathbf{x},f) = \varepsilon_0 \sum_{j,k} \int_{-\infty}^{\infty} df_1\, \chi_{ijk}^{(2)}(-f,f_1,f-f_1)E_j(\mathbf{x},f_1)E_k(\mathbf{x},f-f_1)$$

$$= \varepsilon_0 \sum_{j,k} \int_{-\infty}^{\infty} df_1'\, \chi_{ijk}^{(2)}(-f,f-f_1',f_1')E_j(\mathbf{x},f-f_1')E_k(\mathbf{x},f_1') \tag{1.23}$$

or with $f_1 + f_2 = f_3$:

$$P_i^{(2)}(\mathbf{x},f_3) = \varepsilon_0 \sum_{j,k} \int_{-\infty}^{\infty} df_1\, \chi_{ijk}^{(2)}(-f_3,f_1,f_2)E_j(\mathbf{x},f_1)E_k(\mathbf{x},f_2) \tag{1.24}$$

or symbolic:

$$\mathbf{P}^{(2)}(\mathbf{x},f_3) = \varepsilon_0 \int_{-\infty}^{\infty} \chi^{(2)}(-f_3,f_1,f_2){:}\mathbf{E}(\mathbf{x},f_1)\mathbf{E}(\mathbf{x},f_2)\, df_1 \tag{1.25}$$

where

$$\chi_{ijk}^{(2)}(-f_3,f_1,f_2) = \iint_{-\infty}^{+\infty} \chi_{ijk}^{(2)}(\tau_1,\tau_2)e^{-2\pi jf_1\tau_1}e^{-2\pi jf_2\tau_2}d\tau_1 d\tau_2 \tag{1.26}$$

$$\chi_{ijk}^{(2)}(-f_3,f_1,f_2) = \left[\chi_{ijk}^{(2)}(f_3,-f_1,-f_2)\right]^* \tag{1.27}$$

The first frequency parameter in the argument of $\chi_{ijk}^{(2)}$ is unnecessary in principle; it was added in order that the sum of all three frequency parameters vanishes; at the same time, it is the negative frequency variable of the left-hand-side polarization. This will be also the convention for the third-order susceptibility $\chi_{ijkl}^{(3)}(-f_4,f_1,f_2,f_3)$. Similarly, one should write for the first-order susceptibility $\chi^{(1)}(-f,f)$ instead of the usual $\chi^{(1)}(f)$, but we shall adhere to the simpler notation $\chi(f)$. This notation is not generally accepted in the literature, therefore, one should be careful when reading other texts or articles.

Generalizing Eqs. (1.15) and (1.23), one obtains for the nth-order polarization $P_i^{(n)}$ (spatial dependence suppressed):

$$
P_i^{(n)}(t) = \varepsilon_0 \sum_{i_1\cdots i_n} \int_{-\infty}^{+\infty} \cdots \int \chi_{ii_1\cdots i_n}^{(n)}(\tau_1,\ldots,\tau_n) E_{i_1}(t-\tau_1)\ldots E_{i_n}(t-\tau_n) \prod_{r=1}^{n} d\tau_r
$$

$$(1.28)$$

and with $f = \sum_{r=1}^{n} f_r$:

$$
P_i^{(n)}(f) = \varepsilon_0 \sum_{i_1\cdots i_n} \int_{-\infty}^{+\infty} \cdots \int \chi_{ii_1\cdots i_n}^{(n)}(-f,f_1,\ldots,f_n) E_{i_1}(f_1)\ldots E_{i_n}(f_n) \prod_{r=1}^{n-1} df_r \quad (1.29)
$$

Inserting the expressions (1.14), (1.15), and (1.28) in Eq. (1.4), we find for $\mathbf{P}(\mathbf{x},t)$ a *Volterra series*:

$$
P_i(\mathbf{x},t) = \varepsilon_0 \sum_{j} \int_{-\infty}^{\infty} d\tau\, \chi_{ij}^{(1)}(\tau) E_j(\mathbf{x},t-\tau)
$$

$$
+ \varepsilon_0 \sum_{j,k} \int_{-\infty}^{\infty} d\tau_1 \int_{-\infty}^{\infty} d\tau_2\, \chi_{ijk}^{(2)}(\tau_1,\tau_2) E_j(\mathbf{x},t-\tau_1) E_k(\mathbf{x},t-\tau_2)
$$

$$
+ \varepsilon_0 \sum_{jkm} \int_{-\infty}^{\infty} d\tau_1 \int_{-\infty}^{\infty} d\tau_2 \int_{-\infty}^{\infty} d\tau_3\, \chi_{ijkm}^{(3)}(\tau_1,\tau_2,\tau_3) \times
$$

$$
E_j(t-\tau_1) E_k(t-\tau_2) E_m(t-\tau_3)
$$

$$+\ldots \tag{1.30}$$

as a consequence of the invariance under time displacement.

Whether one has to take into account the nonlinear terms (and up to which order) depends on the ratio of the radiation field to atomic fields, the latter having an order of magnitude of $10^{10}\ldots 10^{11}$ V/m. For small radiation fields (usual light sources have an order of magnitude of about $|E| \sim 10^2$ V/m, sunlight about 600 V/m), the electrons bound to the nucleus are displaced only by about 10^{-18} m. This value is much smaller than atomic distances of about 10^{-10} m (elastic region). However, if we focus a laser beam with 600 W onto an area of $1 \ \mu\text{m}^2$, the displacements are already so large that we must take into account the nonlinearity of the characteristics (deviation of the real potential from a parabolic potential).

Example 1.2   Calculate the electric field $|E|$ for a laser beam with 600 W, focused on an area of $1 \ \mu\text{m}^2$; the refractive index of the material $n = 1.5$. Use Eq. (A.6). ($|E| = 5.5 \times 10^8$ V/m.)   ∎

For solids, the elements of the susceptibility tensor have an order of magnitude of $\chi_{ij}(f) \sim 1$, $\chi_{ijk}(-f_3, f_1, f_2) \sim 10^{-12}$ m/V, $\chi_{ijkl}(-f_4, f_1, f_2, f_3) \sim 10^{-21}$ m^2/V^2. Contrary to our expansion, Eq. (1.30), some authors use the notation $\chi_{ijk}^{(2)} = 2d_{ijk}$ or $= 2d_{ijk}/\varepsilon_0$, and $\chi_{ijkl}^{(3)} = 4\chi_{ijkl}'^{(3)}$ or $= 4\chi_{ijkl}''^{(3)}/\varepsilon_0$. Although we use here the rational MKSA system (SI), the susceptibilities in the literature are frequently quoted in electrostatic CGS units. The conversion between both systems is described in Appendix B.

Example 1.3   An experimental value for InSb in the low-frequency domain $\omega \approx 0$ is $\overline{\chi}^{(2)} = 3.3 \times 10^{-6}$ esu. Calculate the value of $\chi^{(2)}$ in the MKSA system. [$\chi^{(2)} = 4\pi \times 1.1 \times 10^{-10}$ m/V.]   ∎

1.2   SYMMETRY PROPERTIES OF SUSCEPTIBILITY TENSORS

The nth order susceptibility tensor $\chi_{ii_1 \dots i_n}^{(n)}$ is a tensor of rank $(n + 1)$ with 3^{n+1} components. By exploiting symmetries, the number of different and independent components can be reduced.

The following four properties hold quite generally:

The *intrinsic permutation symmetry* is a consequence of the definition, Eq. (1.28):

$$\chi_{ii_1 \dots i_n}^{(n)}(\tau_1, \dots, \tau_n) = P_t \cdot \chi_{ii_1 \dots i_n}^{(n)}(\tau_1, \dots, \tau_n) \tag{1.31}$$

P_t represents any of the $n!$ permutations of the n pairs $(i_1, \tau_1), \dots, (i_n, \tau_n)$:

$$\text{Example:} \quad \chi_{ijkl}^{(3)}(\tau_1, \tau_2, \tau_3) = \chi_{ilkj}^{(3)}(\tau_3, \tau_2, \tau_1)$$

In the frequency domain, this implies the invariance ($f = \sum_r f_r$)

$$\chi_{ii_1 \dots i_n}^{(n)}(-f, f_1, \dots, f_n) = P_t \cdot \chi_{ii_1 \dots i_n}^{(n)}(-f, f_1, \dots, f_n) \tag{1.32}$$

under permutations of the n pairs $(i_1, f_1), \dots, (i_n, f_n)$, with the exclusion of the first pair $(i, -\sum f_r)$:

$$\text{Example:} \quad \chi_{ijk}^{(2)}(-f, f_1, f_2) = \chi_{ikj}^{(2)}(-f, f_2, f_1)$$

From the *reality of susceptibility in the time domain* follows, as already mentioned, in the frequency domain:

$$\left[\chi_{ii_1\ldots i_n}^{(n)}(-f,f_1,\ldots,f_n)\right]^* = \chi_{ii_1\ldots i_n}^{(n)}(f,-f_1,\ldots,-f_n) \tag{1.33}$$

Example: $\left[\chi_{ijk}^{(2)}(-f,f_1,f_2)\right]^* = \chi_{ijk}^{(2)}(f,-f_1,-f_2)$

Causality requires

$$\chi_{ii_1\ldots i_n}^{(n)}(\tau_1,\ldots,\tau_n) = 0 \tag{1.34}$$

as soon as one of the times τ_j becomes negative.

The symmetry properties of the next group of properties are valid only for selected physical systems.

Consider two systems of coordinates (x_1, x_2, x_3) and (x_1', x_2', x_3'). Under orthogonal transformations T (rotations and inversions) between these two systems, a tensor of rank $(n+1)$ transforms as the $(n+1)$-fold product of the coordinates:

$$\chi_{ii_1\ldots i_n}' = \sum T_{ij} T_{i_1 j_1} \ldots T_{i_n j_n} \cdot \chi_{jj_1\ldots j_n}$$

The crystals are divided into 32 crystal classes according to the admissible point groups. Each point group is characterized by some combination of symmetry elements. If the transformation T coincides with a symmetry transformation of the system, then χ remains unchanged: $\chi_{ii_1\ldots i_n}' = \chi_{ii_1\ldots i_n}$. Thus,

$$\chi_{ii_1\ldots i_n} = \sum T_{ij} T_{i_1 j_1} \ldots T_{i_n j_n} \cdot \chi_{jj_1\ldots j_n} \tag{1.35}$$

If one knows all point symmetry transformations of the material (e.g., of the crystal), one can write down a corresponding number of relations, Eq. (1.35).

The most important example is *inversion*. If the inversion $T_{ij} = -\delta_{ij}$ is an admissible symmetry transformation, that is, the crystal is unchanged under the operation of inversion, the crystal then has an inversion center. Therefore, Eq. (1.35) yields for susceptibility tensors of odd rank (n even):

$$\chi_{ii_1\ldots i_n}^{(n)} = -\chi_{ii_1\ldots i_n}^{(n)} = 0 \tag{1.36}$$

That is, in a system with an inversion center, only the tensors of even rank yield nonlinear effects, for example, $\chi_{ijkl}^{(3)}$, but not $\chi_{ijk}^{(2)}$. Eleven crystal classes and the microscopic disordered systems, for example, gases and liquids, possess an inversion center and thus show nonlinear effects only for susceptibility tensors of even rank. The inversion symmetry can be destroyed by spatial inhomogeneities or an applied static electric field.

Appendix C contains a table of the 21 crystal classes without inversion symmetry, and for all 32 crystal classes tables for $\chi_{ij}^{(1)}(f)$, $\chi_{ijk}^{(2)}(-\sum f_r, f_1, f_2)$, $\chi_{ijkl}^{(3)}(-\sum f_r, f_1, f_2, f_3)$.

We assume now up to the end of this subsection *losslessness* of the processes. In the case of losslessness, *invariance under time reversal* holds, that is, the equations of motion (without damping!) are left unchanged if t is replaced by $-t$: The operation time reversal $\mathcal{T}$ causes the transformation

$$G(t) \longrightarrow \mathcal{T}G(t) = G(-t)$$

and invariance under time reversal then requires

$$G(t) = G(-t)$$

Then, by passage to the Fourier transform, one shows easily using Eq. (1.33):

$$\chi_{ii_1...i_n}^{(n)}\left(-\sum f_r, f_1, \ldots, f_n\right) = \chi_{ii_1...i_n}^{(n)*}\left(-\sum f_r, f_1, \ldots, f_n\right): \text{real!} \qquad (1.37)$$

See Problem 1.4. One can see this immediately, for example, for the linear susceptibility tensor, as follows: Losslessness implies that $\varepsilon_{ij}(f)$ is real, and therefore, according to Eq. (1.21), $\chi_{ij}(f)$ is real too!

Moreover, we show in Appendix D that for lossless processes, instead of Eq. (1.32), we have a permutation symmetry for all pairs, the first pair $(i, -\sum f_r)$ being included. This is an *overall permutation symmetry*. For example, one has for $f_3 = f_1 + f_2$:

$$\chi_{ijk}(-f_3, f_1, f_2) = \chi_{jik}(f_1, -f_3, f_2) \qquad (1.38)$$

If the nonlinear medium is lossless throughout a spectral region that includes all the frequencies involved in the interaction, then the permutation symmetry holds for all Cartesian tensor indices alone, for example,

$$\chi_{ijk}(-f_3, f_1, f_2) = \chi_{jik}(-f_3, f_1, f_2) \qquad (1.39)$$

Thus, the 27 elements of the tensor χ_{ijk}, for example, are reduced to only 10 independent elements. This *Kleinman's symmetry relation* is, to be sure, also only an approximation, valid far from resonances, because all media are dispersive on principle. (However, if an absorption band lies in between the frequencies involved, Kleinman's symmetry condition does not hold).

1.3   SURVEY OF NONLINEAR PROCESSES

We are interested only in the cases of quadratic and cubic nonlinearity $\mathbf{P}^{(2)}$, $\mathbf{P}^{(3)}$; see Eq. (1.4). Frequently, one investigates nonlinear optical processes by illumination of matter with stationary fields having very small bandwidths. These fields can approximately be considered monochromatic. If there are fields at p different frequencies $f_1, \ldots, f_p$, we can generalize Eqs. (A.2) and (A.4) to a superposition of p signals of different frequencies:

$$\mathbf{E}(t) = \left(\frac{1}{2}\right) \sum_{r=\pm1}^{\pm p} \hat{\mathbf{E}}(f_r) \cdot e^{j\omega_r t} \tag{1.40}$$

$$\mathbf{E}(f) = \left(\frac{1}{2}\right) \sum_{r=\pm1}^{\pm p} \hat{\mathbf{E}}(f_r) \cdot \delta(f - f_r) \tag{1.41}$$

where $f_{-r} = -f_r$, $\hat{E}(-f_r) = \hat{E}^*(f_r)$. To quadratic polarization only fields with at most two different frequencies f_1, f_2 contribute ($p \leq 2$); to cubic polarization fields with up to three different frequencies f_1, f_2, f_3 ($p \leq 3$); in general, $p \leq n$. All the effects described in this section are treated in detail in Chapters 3 through 8.

We start with quadratic polarization in the case that only fields with a single frequency f_1 appear. We substitute then $E_i(f)$ in the form of Eq. (1.41) with $p = 1$ into Eq. (1.23) and obtain

$$P_i^{(2)}(f) = \left(\frac{1}{2}\right)[\hat{P}_i(2f_1)\delta(f - 2f_1) + \hat{P}_i^*(2f_1)\delta(f + 2f_1)] + \hat{P}_i(0)\delta(f) \tag{1.42}$$

where

$$\hat{P}_i(2f_1) := \frac{1}{2}\varepsilon_0 \sum \chi_{ijk}(-2f_1,f_1,f_1)\hat{E}_j(f_1)\hat{E}_k(f_1) \tag{1.43}$$

$$\hat{P}_i(0) := \frac{1}{2}\varepsilon_0 \sum \chi_{ijk}(0,f_1,-f_1)\hat{E}_j(f_1)\hat{E}_k^*(f_1) \tag{1.44}$$

$\hat{P}_i(2f_1)$ from Eq. (1.43) describes *second harmonic generation* $2f_1$ (SHG) of the fundamental frequency f_1 (*frequency doubling*), $\hat{P}_i(0)$ from Eq. (1.44) *optical rectification* The appearance of both processes can be seen most quickly as follows: We multiply two fields that are both proportional to $\cos \omega_1 t$. Therefore,

$$\cos^2 \omega_1 t = \left(\frac{1 + \cos 2\omega_1 t}{2}\right)$$

and this contains a constant term and a term with twice the frequency, both with the same prefactor.

Both susceptibility tensors $\chi_{ijk}(-2f_1,f_1,f_1)$ and $\chi_{ijk}(0,f_1,-f_1)$ are symmetric in the two last tensor indices.

Proof At first, we immediately have due to Eq. (1.32):

$$\chi_{ijk}(-2f_1,f_1,f_1) = \chi_{ikj}(-2f_1,f_1,f_1) \tag{1.45}$$

Next we obtain, using Eq. (1.37), that is, losslessness, Eqs. (1.33) and (1.32) in this order:

$$\chi_{ijk}(0,f_1,-f_1) = \chi_{ijk}^*(0,f_1,-f_1) = \chi_{ijk}(0,-f_1,f_1) = \chi_{ikj}(0,f_1,-f_1) \quad (1.46)$$

Note that we have (see Appendix A)

$$\widehat{P}_i(0) = \widehat{P}_i^*(0) \quad (1.47)$$

■

More interesting is the case in which two different frequencies $f_1 \neq f_2$ appear in Eq. (1.41) with $p = n = 2$. Insertion into Eq. (1.23) now gives

$$
\begin{aligned}
P_i^{(2)}(f) =& \left(\frac{1}{2}\right)[\widehat{P}_i(f_1+f_2)\delta(f-f_1-f_2) + \widehat{P}_i^*(f_1+f_2)\delta(f+f_1+f_2)] \\
&+ \left(\frac{1}{2}\right)[\widehat{P}_i(f_1-f_2)\delta(f-f_1+f_2) + \widehat{P}_i^*(f_1-f_2)\delta(f+f_1-f_2)] \\
&+ \left(\frac{1}{2}\right)[\widehat{P}_i(2f_1)\delta(f-2f_1) + \widehat{P}_i^*(2f_1)\delta(f+2f_1)] \\
&+ \left(\frac{1}{2}\right)[\widehat{P}_i(2f_2)\delta(f-2f_2) + \widehat{P}_i^*(2f_2)\delta(f+2f_2)] \\
&+ \widehat{P}_i(0)\delta(f)
\end{aligned}
\quad (1.48)
$$

with

$$\widehat{P}_i(f_1+f_2) = \varepsilon_0 \sum \chi_{ijk}(-f_1-f_2,f_1,f_2)\widehat{E}_j(f_1)\widehat{E}_k(f_2) \quad (1.49)$$

$$\widehat{P}_i(f_1-f_2) = \varepsilon_0 \sum \chi_{ijk}(-f_1+f_2,f_1,-f_2)\widehat{E}_j(f_1)\widehat{E}_k^*(f_2) \quad (1.50)$$

$$\widehat{P}_i(2f_1) = \frac{1}{2}\varepsilon_0 \sum \chi_{ijk}(-2f_1,f_1,f_1)\widehat{E}_j(f_1)\widehat{E}_k(f_1) \quad (1.51)$$

$$\widehat{P}_i(2f_2) = \frac{1}{2}\varepsilon_0 \sum \chi_{ijk}(-2f_2,f_2,f_2)\widehat{E}_j(f_2)\widehat{E}_k(f_2) \quad (1.52)$$

$$\widehat{P}_i(0) = \frac{1}{2}\varepsilon_0 \sum [\chi_{ijk}(0,f_1,-f_1)\widehat{E}_j(f_1)\widehat{E}_k^*(f_1)$$

$$+ \chi_{ijk}(0,f_2,-f_2)\widehat{E}_j(f_2)\widehat{E}_k^*(f_2)] \quad (1.53)$$

Summarizing, we write for the different equations, (1.49) through (1.52),

$$\widehat{P}_i^{(2)}(f) = c^{(2)}\varepsilon_0 \sum \chi_{ijk}(-f,f',f'')\widehat{E}_j(f')\widehat{E}_k(f'') \quad (1.54)$$

with $f = f' + f'' \neq 0$ and

$$c^{(2)} = \begin{cases} 1 & \text{for} \quad f'' \neq f' \\ \dfrac{1}{2} & \text{for} \quad f'' = f' \end{cases} \tag{1.55}$$

Note: If we let $f_2 \rightarrow f_1$ in Eq. (1.49) (valid for $f_1 \neq f_2$) we obtain $\widehat{P}_i(2f_1) + \widehat{P}_i(2f_2 = 2f_1)$; similarly, if in Eq. (1.50) $f_2 \rightarrow f_1$, we obtain Eq. (1.53) for $f_2 = f_1$. The results, Eqs. (1.43) and (1.44), are contained in Eqs. (1.51) through (1.53): Eqs. (1.51) and (1.52) correspond exactly to Eq. (1.43) and Eq. (1.53) to (1.44), if we set $\widehat{E}(f_2) = 0$.

The equations (1.49) and (1.50) represent *sum and difference frequency mixing*, Eqs. (1.51) and (1.52) frequency doubling (SHG) either at $2f_1$ or at $2f_2$, and finally, Eq. (1.53) optical rectification. Again, one can see most quickly the formation of sum and difference frequency when two fields with different frequencies occur by looking at the product:

$$\cos\omega_1 t \cdot \cos\omega_2 t = \frac{[\cos(\omega_1 + \omega_2)t + \cos(\omega_1 - \omega_2)t]}{2}$$

In $P_i^{(2)}(f)$ are contained still more effects: If we set in Eqs. (1.49) and (1.50) $f_2 = 0$, we obtain a nonlinear polarization:

$$\widehat{P}_i(f_1) = 2\varepsilon_0 \sum \chi_{ijk}(-f_1, f_1, 0)\widehat{E}_j(f_1)\widehat{E}_k(0) \tag{1.56}$$

This represents the so-called *linear electro-optic effect* (Pockels effect): A dc field (or a low-frequency-modulated field compared with the light frequencies) influences the permittivity at the frequency f_1.

If we replace in Eq. (1.50) $f_1 \rightarrow 2f_1, f_2 \rightarrow f_1$ this leads to

$$\widehat{P}_i(f_1) = \varepsilon_0 \sum \chi_{ijk}(-f_1, 2f_1, -f_1)\widehat{E}_j(2f_1)\widehat{E}_k^*(f_1) \tag{1.57}$$

and the nonlinear interaction between two fields with frequencies $2f_1$ and f_1 can generate a nonlinear polarization at the frequency f_1 via difference frequency mixing.

Parametric effects are also based on $P_i^{(2)}(f)$: A weak signal wave at f_1 and an intense pump wave at f_3 are incident on a nonlinear crystal. The amplification of the wave at f_1 is accompanied by the generation of an idler wave at $f_2 = f_3 - f_1$.

For cubic polarization, one has to use Eq. (1.41) with $p \leq n = 3$. This contains many processes with different frequency combinations. We mention only some of them ($n = 3, p \leq 2$):

$$\widehat{P}_i(3f_1) = \frac{1}{4}\varepsilon_0 \sum \chi_{ijkl}(-3f_1, f_1, f_1, f_1)\widehat{E}_j(f_1)\widehat{E}_k(f_1)\widehat{E}_l(f_1) \tag{1.58}$$

$$\widehat{P}_i(f_1) = \frac{3}{2}\varepsilon_0 \sum \chi_{ijkl}(-f_1, f_2, -f_2, f_1)\widehat{E}_j(f_2)\widehat{E}_k^*(f_2)\widehat{E}_l(f_1) \qquad (1.59)$$

$$\widehat{P}_i(f_1) = \frac{3}{4}\varepsilon_0 \sum \chi_{ijkl}(-f_1, f_1, -f_1, f_1)\widehat{E}_j(f_1)\widehat{E}_k^*(f_1)\widehat{E}_l(f_1) \qquad (1.60)$$

$$\widehat{P}_i(f_1) = 3\varepsilon_0 \sum \chi_{ijkl}(-f_1, f_1, 0, 0)\widehat{E}_j(f_1)\widehat{E}_k(0)\widehat{E}_l(0) \qquad (1.61)$$

For the process $f_4 = f_1 + f_2 + f_3$, we write quite generally:

$$\widehat{P}_i^{(3)}(f_4) = c^{(3)}\varepsilon_0 \sum \chi_{ijkl}(-f_4, f_1, f_2, f_3)\widehat{E}_j(f_1)\widehat{E}_k(f_2)\widehat{E}_l(f_3) \qquad (1.62)$$

Equation (1.58) describes *third harmonic generation* $3f_1$ (THG) of the funda-mental frequency f_1, Eq. (1.59) controls the propagation at the frequency f_1 by the intensity of a wave with the frequency f_2 (see *Raman effect*), Eq. (1.60) describes *self-focusing*: The refractive index seen by a wave with frequency f_1 is larger at those places where its intensity is larger. By the focusing effect, the ray will be contracted to a thread and the medium can be destroyed due to the concentration of the intensity. Finally, Eq. (1.61) describes the *quadratic electro-optic effect* (Kerr effect): The light propagation is influenced by the square of a dc field [or of a low-frequency field; then Eq. (1.59) is used].

The prefactors $c^{(n)}$ in front of ε_0 in the expressions for the nonlinear polarization [see, e.g., Eqs. (1.49) through (1.61)] can easily be calculated in general for an nth-order polarization. We assume that not one of the $(n + 1)$ frequencies $f_1, f_2, \ldots, f_n, \Sigma f_r$ vanishes and (without restricting generality):

- χ is independent of frequency (medium without memory).
- Only one Cartesian component of each P and E will be considered, that is, no tensor indices are required.

Then, the general formula of Eq. (1.28) for $P_i^{(n)}(\mathbf{x}, t)$ reads

$$P^{(n)}(t) = \varepsilon_0 \cdot \chi^{(n)} \cdot [E(t)]^n \qquad (1.63)$$

At first, we insert

$$E(t) = \frac{1}{2}\sum_{s=\pm 1}^{\pm S} \widehat{E}(f_s)\exp(j\omega_s t), \qquad P^{(n)}(t) = \frac{1}{2}\sum_{r=\pm 1}^{\pm R} \widehat{P}^{(n)}(f_r)\exp(j\omega_r t)$$

into Eq. (1.63) and obtain

$$\sum_{r=\pm 1}^{\pm R} \widehat{P}^{(n)}(f_r)\exp(j\omega_r t) = \frac{\varepsilon_0}{2^{n-1}} \cdot \chi^{(n)} \left[\sum_{s=\pm 1}^{\pm S} \widehat{E}(f_s)\exp(j\omega_s t) \right]^n$$

Now we select from the left-hand side a fixed term: the one with $\exp(j\omega_r t)$. With the polynomial theorem follows:

$$\widehat{P}^{(n)}(f_r) = \frac{\varepsilon_0 \chi^{(n)}}{2^{n-1}} {\sum}' \frac{n!}{m_1! m_{-1}! \ldots m_S! m_{-S}!}$$

$$\times [\widehat{E}(f_1)]^{m_1} \cdot [\widehat{E}(f_{-1})]^{m_{-1}} \ldots [\widehat{E}(f_{-S})]^{m_{-S}} \tag{1.64}$$

$\sum'$ means that we have to sum over all terms for which $\sum_s m_s \omega_s = \omega_r$, whereas, for example, the factor m_{-q} states how often the field $\widehat{E}(f_{-q})$ occurs as a factor on the right-hand side; moreover, $\sum_s m_s = n$. For a fixed combination of the $m_1, \ldots, m_{-S}$, the numerical factor at ε_0 then reads

$$c^{(n)} = \frac{1}{2^{n-1}} \cdot \frac{n!}{m_1! m_{-1}! \ldots m_S! m_{-S}!} \tag{1.65}$$

and if all frequencies are $\neq 0$, we have

$$\widehat{P}^{(n)}(f_r) = c^{(n)} \varepsilon_0 \cdot \chi^{(n)}(-f_r, f_1, \ldots, f_n) \widehat{E}(f_1) \ldots \widehat{E}(f_n) \tag{1.66}$$

The formalism can be extended to also include the case with vanishing frequencies. Then (without proof)

$$c^{(n)} = \frac{2^{l+m}}{2^n} \frac{n!}{m_1! m_{-1}! \ldots m_S! m_{-S}!} \tag{1.67}$$

where $l = 1$ if $f_r \neq 0$; otherwise, $l = 0$, and m frequencies out of the set $f_1, f_2, \ldots, f_n$ are zero.

1.4   THE ANHARMONIC OSCILLATOR

Since we dispense with quantum theory in this book, we have to use a classical oscillator model for the calculation of susceptibilities. We consider charge carriers, for example, electrons (charge $q = -e$, mass m), that are bound with springs to fixed space points with charge $+e$, for example, atomic nuclei, and can oscillate in the x-direction. The deviation from the equilibrium position gives rise to a dipole moment. The interaction between the different induced dipole moments will be neglected. For the restoring force of the spring, we use the nonlinear expression:

$$F(x) = m(\omega_0^2 x + a' x^2 + b' x^3 + \ldots)$$

Then, the equation of motion for the forced damped oscillation under the influence of the electric field $E(t)$ reads, after division with m;

$$\ddot{x}(t) + \gamma\dot{x} + \omega_0^2 x = \frac{q}{m} \cdot E(t) - (a'x^2 + b'x^3 + \ldots)$$

ω_0 is the resonance angular frequency and γ a small damping coefficient with $0 < \gamma < 2\omega_0$. In many cases, the coefficient a' vanishes; then $b'x^3$ is the nonlinear term of lowest order. With N as the number of oscillators per unit volume, the polarization—as the volume density of the total dipole moment—becomes $P = Nqx$, and the differential equation for $P(t)$ is

$$\ddot{P}(t) + \gamma\dot{P} + \omega_0^2 P = \frac{Nq^2}{m} \cdot E(t) - \left(\eta a P^2 + \eta^2 b P^3 + \ldots\right) =: s(t) \qquad (1.68)$$

Here, we have set

$$\frac{a'}{(Nq)} = \eta a, \qquad \frac{b'}{(Nq)^2} = \eta^2 b, \qquad \eta \ll 1$$

$s(t)$ is an abbreviation for the right-hand side.

Equation (1.68) is solved iteratively as follows: A Fourier transform for $P(t)$ and $s(t)$:

$$P(t) = \int P(f)e^{j\omega t}df, \qquad s(t) = \int s(f)e^{j\omega t}df \qquad (1.69)$$

leads directly to the solution in the frequency domain:

$$P(f) = \frac{1}{\omega_0^2 + j\gamma\omega - \omega^2} \cdot s(f) = Q(f) \cdot s(f) \qquad (1.70)$$

Note again that $Q^*(f) = Q(-f)$. Then, we have a convolution in the time domain:

$$P(t) = Q(t) * s(t) = \int\limits_{-\infty}^{+\infty} s(t-\tau)Q(\tau)\,d\tau \qquad (1.71)$$

The integral

$$Q(t) = \int\limits_{-\infty}^{+\infty} Q(f)e^{j\omega t}df = -\frac{1}{2\pi}\int\limits_{-\infty}^{+\infty} \frac{e^{j\omega t}d\omega}{\omega^2 - j\gamma\omega - \omega_0^2} \qquad (1.72)$$

can be found easily by integration in the complex plane. The result is

$$Q(t) = \begin{cases} \dfrac{\exp(-\gamma t/2)}{\sqrt{\omega_0^2 - \gamma^2/4}} \sin\left(\sqrt{\omega_0^2 - \gamma^2/4} \cdot t\right) & \text{for} \quad t > 0 \\[2ex] 0 & \text{for} \quad t < 0 \end{cases} \qquad (1.73)$$

Therefore, from Eq. (1.71) (up to terms including η^1), we have

$$P(t) = \int_0^{+\infty} d\tau \, Q(\tau) \left[\frac{Nq^2}{m} E(t - \tau) - \eta a P^2(t - \tau) - \ldots \right] \qquad (1.74)$$

From this, we calculate $P(t - \tau)$ and insert again into the integrand:

$$P(t) = \frac{Nq^2}{m} \int_0^{+\infty} d\tau \, Q(\tau) E(t - \tau)$$

$$- \eta a \int_0^{+\infty} d\tau \, Q(\tau) \left\{ \int_0^{+\infty} d\tau' \, Q(\tau') \left[\frac{Nq^2}{m} E(t - \tau - \tau') - \ldots \right] \right\}^2$$

Comparison with

$$P(t) = P^{(1)}(t) + \eta P^{(2)}(t) + \ldots \qquad (1.75)$$

results in (as convolution)

$$P^{(1)}(t) = \frac{Nq^2}{m} Q(t) * E(t) \qquad (1.76)$$

$$P^{(1)}(f) = \frac{Nq^2}{m} Q(f) \cdot E(f) =: \varepsilon_0 \chi^{(1)}(f) E(f)$$

or

$$\varepsilon_0 \chi^{(1)}(f) = \frac{Nq^2}{m} Q(f) \qquad (1.77)$$

Moreover, we find [with $\tau + \tau' = \tau_1, \tau + \tau'' = \tau_2$, and $M := -a(Nq^2/m)^2$]

$$P^{(2)}(t) = M \int d\tau \, d\tau' \, d\tau'' \, Q(\tau) Q(\tau') Q(\tau'') E(t - \tau - \tau') E(t - \tau - \tau'')$$

$$= M \int d\tau \, d\tau_1 \, d\tau_2 \, Q(\tau) Q(\tau_1 - \tau) Q(\tau_2 - \tau) E(t - \tau_1) E(t - \tau_2)$$

$$=: \varepsilon_0 \int \chi^{(2)}(\tau_1, \tau_2) E(t - \tau_1) E(t - \tau_2) \, d\tau_1 \, d\tau_2$$

Thus,

$$\varepsilon_0 \chi^{(2)}(\tau_1, \tau_2) = -a\left(\frac{Nq^2}{m}\right)^2 \int d\tau \, Q(\tau)Q(\tau_1 - \tau)Q(\tau_2 - \tau) \qquad (1.78)$$

A Fourier transform of this yields on the left-hand side

$$\varepsilon_0 \int \chi^{(2)}(f_1, f_2) \exp(j\omega_1 \tau_1) \exp(j\omega_2 \tau_2) \, df_1 \, df_2$$

and on the right-hand side for the integral

$$\int d\tau \int Q(f)e^{j\omega\tau} df \int Q(f_1)e^{j\omega_1(\tau_1 - \tau)} df_1 \int Q(f_2)e^{j\omega_2(\tau_2 - \tau)} df_2$$

$$= \int df \, df_1 \, df_2 \, Q(f)Q(f_1)Q(f_2) \exp(j\omega_1\tau_1 + j\omega_2\tau_2)\delta(f - f_1 - f_2)$$

$$= \int df_1 df_2 \, Q(f_1 + f_2)Q(f_1)Q(f_2) \exp(j\omega_1\tau_1 + j\omega_2\tau_2)$$

Comparison of the left-hand side with the right one yields, with $f = f_1 + f_2$;

$$\varepsilon_0 \chi^{(2)}(f_1, f_2) = -a\left(\frac{Nq^2}{m}\right)^2 Q(f_1 + f_2)Q(f_1)Q(f_2)$$

$$= -a\frac{m}{Nq^2} \varepsilon_0^3 \chi^{(1)}(f)\chi^{(1)}(f_1)\chi^{(1)}(f_2) \qquad (1.79)$$

$$= \varepsilon_0 \chi^{(2)}(-f, f_1, f_2)$$

meaning high refractory materials should have large nonlinear susceptibilities [large $n \to$ large $\varepsilon \to$ large $\chi^{(1)} \to$ large $\chi^{(2)}$]. This is called Miller's rule, holding true only with restrictions.

We specialize Eq. (1.79) to various processes (with constants $C_1, \ldots, C_4$): Sum frequency mixing $f = f_1 + f_2$:

$$\chi^{(2)}(-f, f_1, f_2) \qquad (1.80)$$

$$= \frac{C_1}{(\omega_0^2 - \omega_1^2 + j\gamma\omega_1)(\omega_0^2 - \omega_2^2 + j\gamma\omega_2)[\omega_0^2 - (\omega_1 + \omega_2)^2 + j\gamma(\omega_1 + \omega_2)]}$$

Difference frequency mixing $f = f_1 - f_2$:

$$\chi^{(2)}(-f, f_1, -f_2) \qquad (1.81)$$

$$= \frac{C_2}{(\omega_0^2 - \omega_1^2 + j\gamma\omega_1)(\omega_0^2 - \omega_2^2 - j\gamma\omega_2)[\omega_0^2 - (\omega_1 - \omega_2)^2 + j\gamma(\omega_1 - \omega_2)]}$$

Frequency doubling (SHG) $f = 2f_1$:

$$\chi^{(2)}(-f, f_1, f_1) = \frac{C_3}{(\omega_0^2 - \omega_1^2 + j\gamma\omega_1)^2 [\omega_0^2 - (2\omega_1)^2 + j\gamma(2\omega_1)]} \quad (1.82)$$

Optical rectification $f = f_1 - f_1$:

$$\chi^{(2)}(0, f_1, -f_1) = \frac{C_4}{[(\omega_0^2 - \omega_1^2)^2 + (\gamma\omega_1)^2] \cdot \omega_0^2} \quad (1.83)$$

These expressions have a structure similar to that of the quantum mechanically derived susceptibilities.

1.5 KRAMERS–KRONIG RELATIONS

In this section, we consider only isotropic media ($\varepsilon_{ij} = \varepsilon \cdot \delta_{ij}$, $\chi_{ij} = \chi \cdot \delta_{ij}$). In Section 1.1, we showed that the linear susceptibility $\chi(f)$ must be a complex quantity. Exactly the same holds for $\varepsilon(f)$, however, the real and imaginary parts are not independent of one another.

According to Eq. (1.16), we have $\chi(\tau) = 0$ for $\tau < 0$ because of causality. Moreover, we assume:

- $\chi(\tau)$ bounded for all $\tau > 0$.
- $\chi(\tau) \to 0$ sufficiently rapidly for $\tau \to \infty$ (finite memory).

Then we have for complex $f' = \operatorname{Re} f' + j \operatorname{Im} f'$, according to Eq. (1.19),

$$\chi(f') = \int_{-\infty}^{\infty} \chi(\tau) e^{-2\pi j\tau \operatorname{Re} f' + 2\pi\tau \operatorname{Im} f'} d\tau = \int_{0}^{\infty} \chi(\tau) e^{-2\pi j\tau \operatorname{Re} f' + 2\pi\tau \operatorname{Im} f'} d\tau \quad (1.84)$$

The integral converges in the lower-frequency half-plane ($\operatorname{Im} f' < 0$) and also for $\operatorname{Im} f' = 0$. Then, $\chi(f')$ has no pole in the lower half-plane with the real f'-axis included. Therefore, according to the Cauchy integral theorem for the specified closed path (see Fig. 1.1),

$$0 = \oint df' \frac{\chi(f')}{f' - f} = \left(\int_C + \int_{C'} + \int_{-R}^{f-\varepsilon} + \int_{f+\varepsilon}^{R} \right) df' \frac{\chi(f')}{f' - f}$$

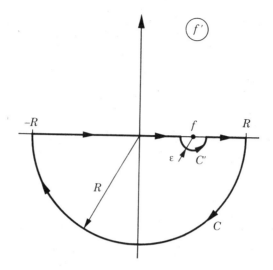

FIGURE 1.1   Closed contour for integration in the complex frequency plane.

The integral over C $[f' = R\exp(j\varphi) = R\cos\varphi + jR\sin\varphi]$ vanishes in the limit $R \to \infty$:

$$\int_C = \int_{2\pi}^{\pi} jRe^{j\varphi}\,d\varphi\,\frac{\chi(Re^{j\varphi})}{R\cdot\exp(j\varphi) - f} = j\int_{2\pi}^{\pi} d\varphi\,\frac{\chi(Re^{j\varphi})}{1 - f\cdot\exp(-j\varphi)/R} \to 0$$

since $\chi(Re^{j\varphi}) \to 0$ due to Eq. (1.84). For the integral over C' $(f' = f + \varepsilon e^{j\varphi})$, we have in the limit $\varepsilon \to 0$:

$$\int_{C'} = \int_{\pi}^{2\pi} j\varepsilon e^{j\varphi}\,\frac{\chi[f + \varepsilon\cdot\exp(j\varphi)]}{\varepsilon\cdot\exp(j\varphi)}\,d\varphi \longrightarrow j\int_{\pi}^{2\pi} d\varphi\chi(f) = \pi j\chi(f)$$

Thus,

$$0 = \pi j\chi(f) - \mathcal{P}\int_{-\infty}^{+\infty} \frac{\chi(f')}{f - f'}\,df'$$

($\mathcal{P}$ means a principal value integral), or

$$\chi(f) = -\frac{j}{\pi}\mathcal{P}\int_{-\infty}^{\infty} \frac{\chi(f')}{f - f'}\,df' \tag{1.85}$$

Decomposition into real and imaginary parts results in

$$\text{Re } \chi(f) = +\frac{1}{\pi} P \int \frac{\text{Im } \chi(f')}{f-f'} df'$$

$$\text{Im } \chi(f) = -\frac{1}{\pi} P \int \frac{\text{Re } \chi(f')}{f-f'} df'$$

(1.86)

These *Kramers–Kronig relations* (*dispersion relations*) determine the connection between the real and imaginary part of linear susceptibility. In mathematics, relations of the form of Eq. (1.86) are called Hilbert transforms.

For the nonlinear susceptibilities—see Eqs. (1.80) through (1.83)—one can also derive dispersion relations in slightly changed form. However, they have found only little application due to their complicated structure (Caspers, 1964, Price, 1963).

PROBLEMS

1.1 Assume $\chi_{ij}(f) = \text{const} = C$. Show that C must be real, and this corresponds to a medium without memory.

1.2 Show that an antisymmetric part of $q_{ijk}(t,\tau_1,\tau_2)$ in Eq. (1.10) makes a vanishing contribution to the polarization.

Section 1.2

1.3 For a rotation about the 3-axis, we have the matrix:

$$(T_{ij}) = \begin{bmatrix} \cos\Theta & -\sin\Theta & 0 \\ \sin\Theta & \cos\Theta & 0 \\ 0 & 0 & 1 \end{bmatrix}$$

The medium is invariant if rotated through $\Theta = 90°$. Derive the relations between the elements χ_{ij} in this case.

1.4 Derive Eq. (1.37).

1.5 Show that $\varepsilon_{ij}(f)$ is a symmetric tensor for a lossless medium.

Section 1.3

1.6 Calculate the prefactors $c^{(2)}$ for the processes $f_1+f_2, f_1-f_2, f_1+f_1, (2f_1)-f_1$ and the prefactors $c^{(3)}$ for $f_1+f_1+f_1, f_1-f_1+f_2, f_1-f_1+f_1$ according to Eq. (1.65).

Section 1.4

1.7 Derive Eq. (1.73).

1.8 Extend the calculations for the anharmonic oscillator up to terms with η^2 (in the case $a = 0$).

Section 1.5

1.9 Because of causality, we have

$$\chi(t) = \chi(t) \cdot H(t)$$

where $H(t)$ is the step function. Use its Fourier transform:

$$H(f) = \mathcal{P}\frac{1}{2\pi jf} + \frac{\delta(f)}{2}$$

to derive the Kramers–Kronig relations (Hu, 1989).

REFERENCES

W. J. Caspers. Dispersion relations for nonlinear response. *Phys. Rev. A.* 133 (1964): 1249–1251.

B. Y. Hu. Kramers–Kronig in two lines. *Am. J. Phys.* 57 (1989): 821.

P. J. Price. Theory of quadratic response functions. *Phys. Rev.* 130 (1963): 1792–1797.

BIBLIOGRAPHY

N. Bloembergen. *Nonlinear Optics.* Addison-Wesley, Reading, MA, 1991.

R. W. Boyd. *Nonlinear Optics.* Academic Press, San Diego, CA, 1992.

P. N. Butcher and D. Cotter. *The Elements of Nonlinear Optics.* Cambridge University Press, Cambridge, UK, 1990.

G. K. Grau. *Quantenelektronik.* Vieweg, Braunschweig, 1978.

A. Nussbaum and R. A. Phillips. *Contemporary Optics.* Prentice-Hall, Englewood Cliffs, NJ, 1976.

B. E. A. Saleh and M. C. Teich. *Fundamentals of Photonics.* Wiley, New York, 1991.

M. Schubert and B. Wilhelmi. *Nonlinear Optics and Quantum Electronics.* Wiley, New York, 1986.

Y. R. Shen. *The Principles of Nonlinear Optics.* Wiley, New York, 1984.

CHAPTER TWO

Wave Propagation in Nonlinear Anisotropic Media

This chapter like the preceding one is basic since all the nonlinear effects in Chapters 3 through 8 depend on Chapters 1 and 2. In the first section, we discuss wave propagation in linear and nonlinear but isotropic media. As an introduction, the Manley–Rowe equations are derived and the connection to energy conservation is shown. Then, the differential equations that describe the nonlinear processes are derived and the slowly varying amplitude (or envelope) approximation is introduced. A short discussion of the connection between linear susceptibility and the propagation vector closes this first section.

In the second section, we consider anisotropic media. At first, we derive the properties of light propagation in these media and use these properties in discussing phase matching, a method for improving the energy exchange between different waves. This phase matching is connected to the conservation of momentum.

2.1 WAVES IN LINEAR AND NONLINEAR MEDIA

2.1.1 Power Balance, Manley–Rowe Equations, and Energy Conservation

From Maxwell's equations,

$$\nabla \times \mathbf{E}(\mathbf{x}, t) = -\dot{\mathbf{B}} = -\mu_0 \dot{\mathbf{H}}(\mathbf{x}, t)$$
$$\nabla \times \mathbf{H}(\mathbf{x}, t) = \dot{\mathbf{D}}(\mathbf{x}, t) + \boldsymbol{\sigma} \cdot \mathbf{E}(\mathbf{x}, t) = \varepsilon_0 \dot{\mathbf{E}}(\mathbf{x}, t) + \dot{\mathbf{P}}(\mathbf{x}, t) + \boldsymbol{\sigma} \cdot \mathbf{E}(\mathbf{x}, t) \quad (2.1)$$

[$\boldsymbol{\sigma}$ = conductivity tensor; the ith component of the vector $\boldsymbol{\sigma} \cdot \mathbf{E}(\mathbf{x}, t)$ is $\sum_j \sigma_{ij} E_j$], we obtain

$$-\nabla \cdot (\mathbf{E} \times \mathbf{H}) = \left(\frac{1}{2}\right) \partial_t (\varepsilon_0 \mathbf{E}^2 + \mu_0 \mathbf{H}^2) + \mathbf{E} \cdot \partial_t \mathbf{P} + \boldsymbol{\sigma}: \mathbf{EE} \quad (2.2)$$

This equation represents power balance. The electromagnetic power density consumed (or gained) at time t at position $\mathbf{x}$, $-\nabla \cdot (\mathbf{E} \times \mathbf{H}) = -\nabla \cdot \mathbf{S}$, with $\mathbf{S}$ as the Poynting vector, is distributed as follows:

- $\boldsymbol{\sigma}: \mathbf{EE} = \sum_{i,j} \sigma_{ij} E_i E_j$ describes losses from conductivity.
- $\mathbf{E} \cdot \dot{\mathbf{P}} \gtrless 0$ is the power density consumed or gained from the medium due to polarization.
- $(1/2) \cdot \partial_t(\ldots)$ is the temporal change of energy density of the electromagnetic field (in vacuum).

The time-averaged dielectric losses or gains due to polarization are by definition

$$p_\ell = \langle \mathbf{E} \cdot \dot{\mathbf{P}} \rangle_t \tag{2.3}$$

If there are many monochromatic waves at the frequencies f_r, we have for the losses

$$p_\ell = \sum_r p_{\ell r} \tag{2.4}$$

where—by using Eq. (A.2)—the loss $p_{\ell r}$ at the fixed frequency f_r is calculated as

$$p_{\ell r} = \frac{j\omega_r}{4} \cdot \sum_i \left[\widehat{E}_i^*(f_r)\widehat{P}_i(f_r) - \widehat{E}_i(f_r)\widehat{P}_i^*(f_r) \right] \tag{2.5}$$

since the terms with $\exp(\pm 2\omega_r t)$ vanish when time-averaged.

For the derivation of the Manley–Rowe equations, we consider for simplicity only a process with second-order nonlinearity with three interacting waves at the frequencies f_1, f_2, f_3, for which $f_3 = f_1 + f_2$ $(f_1 \neq f_2)$ holds. Then the polarizations

$$\widehat{P}_i(f_3) = \varepsilon_0 \sum \chi_{ijk}(-f_3, f_1, f_2)\widehat{E}_j(f_1)\widehat{E}_k(f_2)$$

$$\widehat{P}_i(f_1) = \varepsilon_0 \sum \chi_{ijk}(-f_1, -f_2, f_3)\widehat{E}_j^*(f_2)\widehat{E}_k(f_3) \tag{2.6}$$

$$\widehat{P}_i(f_2) = \varepsilon_0 \sum \chi_{ijk}(-f_2, -f_1, f_3)\widehat{E}_j^*(f_1)\widehat{E}_k(f_3)$$

are required; note that $f_1 = -f_2 + f_3$ and $f_2 = -f_1 + f_3$. Hence, we can calculate the following quantities:

$$M_1 := \frac{p_{\ell 1}}{\omega_1} = \frac{j\varepsilon_0}{4} \sum [\chi_{ijk}(-f_1, -f_2, f_3)\widehat{E}_i^*(f_1)\widehat{E}_j^*(f_2)\widehat{E}_k(f_3)$$

$$- \chi_{ijk}^*(-f_1, -f_2, f_3)\widehat{E}_i(f_1)\widehat{E}_j(f_2)\widehat{E}_k^*(f_3)]$$

$$M_2 := \frac{p_{\ell 2}}{\omega_2} = \frac{j\varepsilon_0}{4} \sum [\chi_{ijk}(-f_2, -f_1, f_3)\widehat{E}_i^*(f_2)\widehat{E}_j^*(f_1)\widehat{E}_k(f_3) - \text{c.c.}]$$

$$= \frac{j\varepsilon_0}{4} \sum [\chi_{jik}(-f_2, -f_1, f_3)\widehat{E}_i^*(f_1)\widehat{E}_j^*(f_2)\widehat{E}_k(f_3) - \text{c.c.}] \quad (2.7)$$

$$M_3 := \frac{p_{\ell 3}}{\omega_3} = \frac{j\varepsilon_0}{4} \sum [\chi_{ijk}(-f_3, f_1, f_2)\widehat{E}_i^*(f_3)\widehat{E}_j(f_1)\widehat{E}_k(f_2) - \text{c.c.}]$$

$$= \frac{j\varepsilon_0}{4} \sum [\chi_{kij}(-f_3, f_1, f_2)\widehat{E}_i(f_1)\widehat{E}_j(f_2)\widehat{E}_k^*(f_3) - \text{c.c.}]$$

Taking into account Eqs. (1.33) and (1.38)—for lossless processes ($p_\ell = 0$)—we obtain immediately

$$M_1 = M_2 = -M_3, \qquad \frac{p_{\ell 1}}{\hbar\omega_1} = \frac{p_{\ell 2}}{\hbar\omega_2} = -\frac{p_{\ell 3}}{\hbar\omega_3} \qquad (2.8)$$

These are the *Manley–Rowe equations*. But we also can write

$$p_{\ell r} = \Delta N_r \cdot \hbar\omega_r \qquad (2.9)$$

with ΔN_r as the number of photons with frequency f_r, which are generated or annihilated per unit volume and per unit time by the nonlinear optical interaction. In the photon picture, one can then interpret Eq. (2.8) as follows: Together with one photon of frequency f_1, one photon of frequency f_2 is annihilated; simultaneously, one photon of frequency $f_3 = f_1 + f_2$ is generated or vice versa. Thus, the total photon number is not conserved. For a lossless interaction in the nonlinear medium, one obtains a gain in the electromagnetic power at some frequency only at the cost of losses at some other frequencies.

The Manley–Rowe equations can also be derived from energy conservation. For the participating photons, we have

$$\hbar\omega_1 + \hbar\omega_2 = \hbar\omega_3$$

Thus,

$$0 = (\hbar\omega_1 + \hbar\omega_2 - \hbar\omega_3) \cdot \frac{p_{\ell 1}}{\hbar\omega_1} = p_{\ell 1} + \frac{\hbar\omega_2}{\hbar\omega_1} \cdot p_{\ell 1} - \frac{\hbar\omega_3}{\hbar\omega_1} \cdot p_{\ell 1}$$

But for lossless processes, we get

$$0 = p_\ell = p_{\ell 1} + p_{\ell 2} + p_{\ell 3} \qquad (2.10)$$

Comparison of both expressions leads exactly to Eqs (2.8).

The Manley–Rowe equations (2.8) can be generalized (without proof). Two original frequencies ω_1 and ω_2 can generate waves at the combination frequencies

$$\omega_{mn} = m\omega_1 + n\omega_2 \qquad (2.11)$$

$(m, n = 0, \pm 1, \pm 2, \ldots)$. We then have

$$\sum_{m,n} \frac{n}{\omega_{mn}} \cdot p_{\ell(mn)} = \sum_{m,n} \frac{m}{\omega_{mn}} \cdot p_{\ell(mn)} = 0 \qquad (2.12)$$

Example 2.1   Apply the generalized Manley–Rowe equations to $\omega_3 = \omega_1 + \omega_2$.
The pair $(m, n) = (1, 0)$ in Eq. (2.11) yields $\omega_{10} = \omega_1$, the pair $(0, 1)$ gives $\omega_{01} = \omega_2$, and the pair $(1, 1)$ yields $\omega_{11} = \omega_3$. From Eq. (2.12), we then find $p_{\ell(01)}/\omega_{01} + p_{\ell(11)}/\omega_{11} = 0$ and $p_{\ell(10)}/\omega_{10} + p_{\ell(11)}/\omega_{11} = 0$. However, this agrees with Eq. (2.8).   ∎

2.1.2 System of Differential Equations for the Complex Phasors

Elimination of **H** from Maxwell's equations yields

$$\nabla \times \nabla \times \mathbf{E}(\mathbf{x}, t) = -\mu_0(\ddot{\mathbf{D}} + \boldsymbol{\sigma} \cdot \dot{\mathbf{E}}) = -\mu_0 \boldsymbol{\sigma} \cdot \dot{\mathbf{E}}(\mathbf{x}, t) - \mu_0[\varepsilon_0 \ddot{\mathbf{E}}(\mathbf{x}, t) + \ddot{\mathbf{P}}(\mathbf{x}, t)] \qquad (2.13)$$

or, in the frequency domain, for the Fourier spectra

$$[(\nabla \times \nabla \times) - \mu_0 \varepsilon_0 \omega^2 + j\mu_0 \omega \boldsymbol{\sigma} \cdot] \mathbf{E}(\mathbf{x}, f) = \mu_0 \omega^2 [\mathbf{P}_L(\mathbf{x}, f) + \mathbf{P}_{NL}(\mathbf{x}, f)] \qquad (2.14)$$

Considering monochromatic signals at the frequency f_0, we obtain for the complex phasors

$$[(\nabla \times \nabla \times) - \mu_0 \varepsilon_0 \omega_0^2 + j\mu_0 \omega_0 \boldsymbol{\sigma} \cdot] \widehat{\mathbf{E}}(\mathbf{x}, f_0) = \mu_0 \omega_0^2 [\widehat{\mathbf{P}}_L(\mathbf{x}, f_0) + \widehat{\mathbf{P}}_{NL}(\mathbf{x}, f_0)] \qquad (2.15)$$

If the field contains several different frequencies, we have such an Eq. (2.15) for each frequency. However, a nonlinear polarization depends nonlinearly on the complex phasors of the fields at the various frequencies. Thus, the equations become coupled, and we obtain a coupled nonlinear system of differential equations for the complex phasors of the fields at all participating frequencies.

For linear polarization, we can write according to Eq. (1.18)

$$\widehat{\mathbf{P}}_L = \varepsilon_0 \boldsymbol{\chi} \cdot \widehat{\mathbf{E}}(\mathbf{x}, f) = \varepsilon_0[\boldsymbol{\chi}_{LC}(f) + \boldsymbol{\chi}_{LA}(f)] \cdot \widehat{\mathbf{E}}(\mathbf{x}, f) \qquad (2.16)$$

That is, one wants to treat very often separately different causes of polarization; for example, the susceptibility of a dielectric crystal (χ_{LC}) may be almost constant in the considered frequency domain. However, by doping the material with a relatively small number of atoms of a certain kind, one finds a strongly frequency-dependent component (χ_{LA}). We then combine

$$\boldsymbol{\varepsilon} = \boldsymbol{\varepsilon}(f) := \varepsilon_0[\mathbf{I} + \boldsymbol{\chi}_{LC}(f)] \qquad (2.17)$$

and this is approximately constant and real. Then, Eq. (2.15) transforms into

$$\left[(\nabla \times \nabla \times) - \mu_0\omega_0^2\varepsilon + j\mu_0\omega_0\sigma\right]\widehat{\mathbf{E}}(\mathbf{x},f_0) = \mu_0\omega_0^2(\varepsilon_0\chi_{LA}\cdot\widehat{\mathbf{E}} + \widehat{\mathbf{P}}_{NL})\qquad(2.18)$$

If one is calculating polarization from the microstructure, one has to take into account the microscopic fields.

2.1.3 Approximation of the Slowly Varying Amplitude

The coupling of different waves in a nonlinear medium results in an energy exchange between these waves, for example, the magnitudes of the wave amplitudes can change along the direction $\mathbf{e}_z$ of propagation. To describe this, we derive approximately valid differential equations. They are basic for treating nonlinear processes.

We assume an isotropic medium without Joule's losses and χ_{LA}:

$$\varepsilon_{ij} = \varepsilon \cdot \delta_{ij}, \qquad \sigma_{ij} = 0, \qquad \chi_{LA} = 0$$

Moreover, we set

$$\nabla \times \nabla \times \mathbf{E} = -\nabla^2\mathbf{E}, \qquad \nabla \cdot \mathbf{E} = 0$$

This is approximately fulfilled, for example, in media whose refractive index depends only weakly on the coordinates. We consider a monochromatic linearly polarized wave (at a fixed frequency f) of the form

$$\mathbf{E}(z,t) = \mathbf{e}(f)|\bar{E}(z,f)| \cdot \cos[\omega t - kz + \arg\bar{E}(z,f)]$$

$$= \frac{\mathbf{e}(f)}{2}|\bar{E}(z,f)| \cdot \{\exp j[\omega t - kz + \arg\bar{E}(z,f)] + \text{c.c.}\}$$

$$= \frac{\mathbf{e}(f)}{2}[\bar{E}(z,f)\exp j(\omega t - kz) + \text{c.c.}]$$

with $\mathbf{e}(f)$ as the real dimensionless polarization unit vector:

$$\mathbf{e}^2(f) = \sum e_i(f)e_i(f) = 1, \qquad e_i(f) = e_i^*(f) = e_i(-f)$$

and $k^2(f) = \omega^2\mu_0\varepsilon(f)$. In analogy to Eqs. (A.2) and (A.5), we take as the complex analytic signal and complex phasor, respectively,

$$\widetilde{\mathbf{E}}(z,t) = \widehat{\mathbf{E}}(z,f)e^{+j\omega t}, \qquad \widehat{\mathbf{E}}(z,f) = \mathbf{e}(f)\cdot\bar{E}(z,f)\cdot e^{-jk(f)z}\qquad(2.19)$$

If $\bar{E}(z,f)$ denotes a function only weakly dependent on z ($' =$ derivation with respect to z) with

$$|\bar{E}''| \ll |k\bar{E}'|\qquad(2.20)$$

we obtain the *slowly varying amplitude* or *envelope approximation* (SVAA or SVEA). From Eq. (2.18), we obtain if substituting f_0 by f

$$-\widehat{E}_i''(z,f) = \mu_0\omega^2\varepsilon(f)\widehat{E}_i(z,f) + \mu_0\omega^2\widehat{P}_{NL,i}(z,f) = k^2(f)\widehat{E}_i + \mu_0\omega^2\widehat{P}_{NL,i}$$

According to Eq. (2.19), the expression on the left-hand side yields explicitly

$$-\frac{\partial^2}{\partial z^2}\widehat{E}_i = e_i e^{-jkz}(2jk\bar{E}' + k^2\bar{E} - \bar{E}'') \approx e_i e^{-jkz}(2jk\bar{E}' + k^2\bar{E})$$

because of Eq. (2.20). Thus, we get finally as ODE in the SVAA

$$\bar{E}'(z,f) = \frac{\omega^2 \cdot \mu_0}{2jk}\sum_i \widehat{P}_{NL,i}e_i \cdot e^{jkz} = \frac{\omega^2 \cdot \mu_0}{2jk}\widehat{\mathbf{P}}_{NL} \cdot \mathbf{e}\, e^{jkz} \qquad (2.21)$$

For a second-order polarization, we write according to Eq. (1.54) for $f \neq 0$

$$\widehat{P}_i^{(2)}(z,f) = c^{(2)}\varepsilon_0 \sum \chi_{ijk}(-f,f',f-f')\widehat{E}_j(z,f')\widehat{E}_k(z,f-f') \qquad (2.22)$$

and obtain, with $\Delta k = k(f) - k(f') - k(f-f')$,

$$\bar{E}'(z,f) = -c^{(2)}j\frac{\omega^2 \cdot \mu_0\varepsilon_0}{2k(f)}\sum_{ijk}\chi_{ijk}e_i(f)e_j(f')e_k(f-f')\bar{E}(f')\bar{E}(f-f')e^{j\Delta kz}$$

$$(2.23)$$

or

$$\bar{E}'(z,f) = -j\frac{\omega\chi_{\text{eff}}(-f,f',f-f')}{2cn(f)} \cdot \bar{E}(z,f')\bar{E}(z,f-f')e^{j\Delta kz} \qquad (2.24)$$

if we introduce an effective susceptibility for linearly polarized fields of the form:

$$\chi_{\text{eff}}(-f,f',f-f') = c^{(2)}\sum\chi_{ijk}(-f,f',f-f')e_i(f)e_j(f')e_k(f-f') \qquad (2.25)$$

with

$$c^{(2)} = \begin{cases} 1 & \text{for} \quad f \neq 0, f \neq 2f' \\ \dfrac{1}{2} & \text{for} \quad f = 0, f = 2f' \end{cases}$$

The factor $1/2$ at $f = 2f'$ results from Eq. (1.54) and (1.55). The case $f = 0$

follows from Eq. (1.53) with $\widehat{E}(f_2) = 0$; see Eq. (1.44). It easily can be shown that for lossless processes

$$\chi_{\text{eff}}(-2f_1, f_1, f_1) = \left(\frac{1}{2}\right) \cdot \chi_{\text{eff}}(-f_1, 2f_1, -f_1) \tag{2.26}$$

holds.

Equation (2.21), or (2.24), represents a set of differential equations that are fundamental for the calculation of all nonlinear processes in Chapters 4 through 8. In principle, one can still allow for $\widehat{\mathbf{E}}$ in Eq. (2.19) a weak time dependence. Then, one can treat nonstationary problems too.

2.1.4   Linear Susceptibility and Propagation Vector

We assume again an isotropic medium:

$$\varepsilon_{ij}(f) = \varepsilon\delta_{ij}, \quad \sigma_{ij} = \sigma\delta_{ij}, \quad \chi_{ij}(f) = \chi(f)\delta_{ij} = (\chi' + j\chi'')\delta_{ij}$$

For a plane wave propagating in the z-direction $(\mathbf{k} = k \cdot \mathbf{e}_z)$ and linearly polarized in the x-direction, we set

$$E_x(z, t) = \text{Re } \tilde{E}(z, t) = \text{Re } \widehat{E}(z, f)e^{j\omega t} = \text{Re } \bar{E}e^{j(\omega t - kz)} \tag{2.27}$$

Insertion into Eq. (2.15) yields if the nonlinear polarization P_{NL} is neglected

$$k^2 - \mu_0\omega^2\varepsilon + j\omega\mu_0\sigma = \mu_0\varepsilon_0\omega^2\chi(f)$$

With

$$k_0 = \frac{\omega}{c}$$

$$n = \sqrt{\varepsilon_r} = \sqrt{\frac{\varepsilon}{\varepsilon_0}} = c\sqrt{\varepsilon\mu_0}$$

and the impedance $Z_0 = \sqrt{\mu_0/\varepsilon_0}$ of the vacuum, we find

$$
\begin{aligned}
k = k' + jk'' = k_0 n_{\text{eff}} \quad &= k_0(n'_{\text{eff}} + jn''_{\text{eff}}) \\
= k_0 n\sqrt{1 + \frac{\chi' + j\chi''}{n^2} - j\frac{\sigma Z_0}{k_0 n^2}} \quad &\approx k_0\left[n\left(1 + \frac{\chi'}{2n^2}\right) + j\left(\frac{\chi''}{2n} - \frac{\sigma Z_0}{2nk_0}\right)\right]
\end{aligned} \tag{2.28}
$$

The approximation holds for small values of the conductivity σ and susceptibility χ. Because of $\exp(-jkz) = \exp(-jk'z)\exp(k''z)$, the linear susceptibility

$\chi = \chi' + j\chi''$ accounts for the refractive index in the form $n(1 + \chi'/2n^2)$ and the damping constant $\alpha = -k''$ as

$$\alpha = \frac{(\sigma Z_0 - k_0 \chi'')}{2n} \tag{2.29}$$

2.2   WAVE PROPAGATION IN ANISOTROPIC MEDIA

2.2.1   Fresnel's Equation and Index Ellipsoid

We consider linear problems in lossless media. Then the real symmetric permittivity tensor in Eq. (1.20):

$$D_i(\mathbf{x},f) = \sum \varepsilon_{ij}(f) E_j(\mathbf{x},f)$$

can be diagonalized by a rotation of the system of coordinates. In this system of *principal dielectric axes*, we have

$$\varepsilon_{ij}(f) = \varepsilon_{Hi}\delta_{ij}, \qquad D_i(\mathbf{x},f) = \varepsilon_{Hi}E_i(\mathbf{x},f) \tag{2.30}$$

Then

$$n_{Hi} = \sqrt{\frac{\varepsilon_{Hi}}{\varepsilon_0}} \tag{2.31}$$

states the connection between the *principal refractive indices* n_{Hi} and the *principal dielectric constants* ε_{Hi}. $\mathbf{D}$ and $\mathbf{E}$ are parallel only if $\mathbf{E}$ points in the direction of one of these principal dielectric axes, but not in general.

Introducing in Maxwell's equations (2.1) for the fields $\mathbf{E}$ and $\mathbf{H}$ the ansatz ($\mathbf{F} = \mathbf{E}$ or $\mathbf{H}$)

$$\mathbf{F}(\mathbf{x},t) = \mathrm{Re} \ \hat{\mathbf{F}}_0(\mathbf{x},f) \ \exp(j\omega t) = \mathrm{Re} \ \overline{\mathbf{F}}_0(f) \exp j(\omega t - \mathbf{kx}) \tag{2.32}$$

of a plane monochromatic wave, we find with $\sigma_{ij} = 0$

$$\begin{aligned} \mathbf{k} \times \overline{\mathbf{E}}_0 &= \mu_0 \omega \overline{\mathbf{H}}_0 \\ -\mathbf{k} \times \overline{\mathbf{H}}_0 &= \omega \overline{\mathbf{D}}_0 \end{aligned} \tag{2.33}$$

Thus, there are two systems of orthogonal vector triplets: $(\mathbf{D}_0, \mathbf{H}_0, \mathbf{k})$ and $(\mathbf{E}_0, \mathbf{H}_0, \mathbf{S})$, and $\mathbf{D}_0, \mathbf{E}_0, \mathbf{k}$, and $\mathbf{S}$ lie in a single plane $\perp \mathbf{H}_0$. Therefore, in a crystal the direction of energy propagation, $\mathbf{S} = \mathbf{E} \times \mathbf{H}$, does not in general coincide with the direction of the wave normal $\mathbf{k}$.

Elimination of $\overline{\mathbf{H}}_0$ from Eq. (2.33) and using $\mathbf{k} = s k = s k_0 n$ ($s^2 = 1$) yield the expression:

$$-\mathbf{k} \times (\mathbf{k} \times \overline{\mathbf{E}}_0) = \mathbf{k} \times (\overline{\mathbf{E}}_0 \times \mathbf{k}) = k_0^2 n^2 [\overline{\mathbf{E}}_0 - \mathbf{s}(\mathbf{s} \cdot \overline{\mathbf{E}}_0)] = \mu_0 \omega^2 \overline{\mathbf{D}}_0 \qquad (2.34)$$

In the system of the principal dielectric axes, we get [for the ith vector component and with $k_0^2/(\mu_0 \omega^2) = \varepsilon_0$]

$$\varepsilon_0 n^2 [\overline{E}_{0i} - s_i(\mathbf{s} \cdot \overline{\mathbf{E}}_0)] = \varepsilon_{Hi} \overline{E}_{0i} = \varepsilon_0 n_{Hi}^2 \overline{E}_{0i} \qquad (2.35)$$

or

$$\overline{E}_{0i} = \frac{n^2 s_i (\mathbf{s} \cdot \overline{\mathbf{E}}_0)}{n^2 - n_{Hi}^2}$$

Hence, we now construct the expression $s_1 \overline{E}_{01} + s_2 \overline{E}_{02} + s_3 \overline{E}_{03} = \mathbf{s} \cdot \overline{\mathbf{E}}_0$ and get

$$1 = \sum_{i=1}^{3} \frac{n^2 s_i^2}{n^2 - n_{Hi}^2} = \sum_{i=1}^{3} \frac{s_i^2}{1 - (n_{Hi}/n)^2} \qquad (2.36)$$

Using $1 = \mathbf{s}^2 = s_1^2 + s_2^2 + s_3^2$ on the left-hand side of Eq. (2.36), we obtain

$$\frac{s_1^2}{1 - (n/n_{H1})^2} + \frac{s_2^2}{1 - (n/n_{H2})^2} + \frac{s_3^2}{1 - (n/n_{H3})^2} = 0 \qquad (2.37)$$

This is *Fresnel's equation* for the wave normal: For a given direction $\mathbf{s}$ of the wave normal, this represents a biquadratic equation for n, that is, to each direction $\mathbf{s}$, there are two, (positive) values for n: $n_{(1)}$ and $n_{(2)}$. Thus, there are also two phase velocities c/n. For each of these n-values, the homogeneous system of equations (2.35) is nontrivially solvable (determinant of coefficients = 0) and this results in two ratios $\overline{E}_{01} : \overline{E}_{02} : \overline{E}_{03}$ and correspondingly two other ratios $\overline{D}_{01}: \overline{D}_{02}: \overline{D}_{03}$. These two eigenwaves, $\mathbf{D}_{(1)}$ and $\mathbf{D}_{(2)}$, are linearly polarized (since these ratios are real) with orthogonal polarizations (see Problems 2.5 and 2.6).

Equation (2.2) with $\sigma = 0$ can be written in the linear case in the form:

$$-\boldsymbol{\nabla} \cdot [\mathbf{E} \times \mathbf{H}] = \left(\frac{1}{2}\right) \partial_t (\mathbf{E} \cdot \mathbf{D} + \mu_0 \mathbf{H}^2) = \partial_t (w_{\text{el}} + w_{\text{mg}})$$

with the energy density of the electric part of the field (lossless case, ε_{ij} real and symmetric):

$$w_{\text{el}} = \left(\frac{1}{2}\right) \sum \varepsilon_{ij} E_i E_j \qquad (2.38)$$

(similarly w_{mg}). In the system of principal axes, we get from this

$$2w_{\mathrm{el}} = \varepsilon_{H1}E_1^2 + \varepsilon_{H2}E_2^2 + \varepsilon_{H3}E_3^2 = \frac{D_1^2}{\varepsilon_{H1}} + \frac{D_2^2}{\varepsilon_{H2}} + \frac{D_3^2}{\varepsilon_{H3}}$$

and such represents a general ellipsoid, which after a suitable normalization $[x_i = D_i(2\varepsilon_0 w_{el})^{-1/2}]$ can be written in the form:

$$\frac{x_1^2}{n_{H1}^2} + \frac{x_2^2}{n_{H2}^2} + \frac{x_3^2}{n_{H3}^2} = 1 \tag{2.39}$$

This is the *index ellipsoid* (optical indicatrix, ellipsoid of the wave normal).

There exist now the optically isotropic media with $\varepsilon_{H1} = \varepsilon_{H2} = \varepsilon_{H3}$ (cubic crystals), the optically uniaxial media with $\varepsilon_{H1} = \varepsilon_{H2} \neq \varepsilon_{H3}$ (tetragonal,

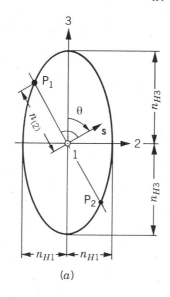

(a)

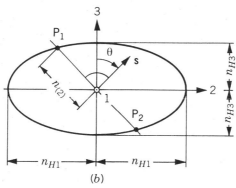

(b)

FIGURE 2.1   Index ellipsoid of optically uniaxial media (rotationally symmetric about the 3-axis): (*a*) positively uniaxial and (*b*) negatively uniaxial.

trigonal, and hexagonal crystals), and the optically biaxial media, where all three ε_{Hi} are different (triclinic, monoclinic, and orthorhombic crystals). Optically biaxial media will not be considered in this book. For the optically uniaxial crystals, the index ellipsoid becomes a spheroid (ellipsoid of revolution) with a distinguished 3-axis (= optical axis). If this spheroid is prolate (= cigarlike), $n_{H3} > n_{H1} = n_{H2}$, the medium is called positively uniaxial; if the spheroid is oblate (= disklike), $n_{H3} < n_{H1} = n_{H2}$, the medium is called negatively uniaxial (see Fig. 2.1). For example, at the wavelength of the Na-D spectralline ($\lambda = 0.589$ μm), one has for quartz $n_{H3} = 1.5553$, $n_{H1} = 1.5442$, whereas for calcite $n_{H3} = 1.4864$, $n_{H1} = 1.6584$ (also see Problem 2.10).

With a geometric construction, we can find the direction of the two eigenwaves $\mathbf{D}_{(1)}$ and $\mathbf{D}_{(2)}$. The index ellipsoid will be intersected with a plane through the origin, normal to the given direction $\mathbf{s}$:

$$\mathbf{s} \cdot \mathbf{x} = s_1 x_1 + s_2 x_2 + s_3 x_3 = 0$$

The line of intersection is an ellipse. The half-lengths of the principal axes of this ellipse are just $n_{(1)}$, $n_{(2)}$, and their directions are those of $\mathbf{D}_{(1)}$, $\mathbf{D}_{(2)}$; the vectors $\mathbf{E}_{(i)}$ are normal to the corresponding tangential planes (without proof).

If the wave vector points in the 3-direction, $\mathbf{s} = \pm \mathbf{e}_3$, then the line of intersection is a circle: The principal axes of the ellipse can be chosen freely (thus also the direction of polarization) and one has $n_{(1)} = n_{(2)} = n_{H1}$. If the angle between the vector $\mathbf{s}$ and the 3-axis is $\theta \neq 0$, we fix the directions of $\mathbf{D}_{(1)}$ and $\mathbf{D}_{(2)}$ as follows:

1. $\mathbf{D}_{(1)}$ is normal to the plane of Fig. 2.1, spanned by $\mathbf{s}$ and the optical axis (3-direction). The pertaining index of refraction is $n_{(1)} = n_{H1} = n_o$, independent of θ. This represents the "ordinary" wave.

2. $\mathbf{D}_{(2)}$ lies in the plane of the figure; for the refractive index $n_{(2)}$ pertaining to this direction of polarization, one obtains

$$\frac{1}{n_{(2)}^2} = \frac{\cos^2 \theta}{n_{H1}^2} + \frac{\sin^2 \theta}{n_{H3}^2} \tag{2.40}$$

or

$$\tan \theta = \frac{n_{H3}}{n_{H1}} \sqrt{\frac{n_{(2)}^2 - n_{H1}^2}{n_{H3}^2 - n_{(2)}^2}} \tag{2.41}$$

that is, this index of refraction $n_{(2)} = n_e$ is dependent on the direction of propagation $\mathbf{s}$; this is the "extraordinary" wave.

2.2.2   Phase Matching and Conservation of the Momentum

Between two waves propagating synchronously, that is, with equal phase velocities in the same direction, we expect a very intense cumulative interaction.

In nonlinear optics, we frequently require synchronism for waves with different frequencies; this requirement can be fulfilled in anisotropic media by the so-called phase-matching.

In particular, we look for a common direction $\mathbf{s}$ of the wave normals in an optical uniaxial medium, for which a wave of frequency f_1 has the same phase velocity (i.e., the same refractive index) as a wave at frequency $f_3 = f_1 + f_2 = 2f_1$ (for the special case $f_2 = f_1$). Figure 2.2 shows the index ellipsoids of a positively uniaxial medium at the frequencies $2f_1$ and f_1. *Note*: In normal dispersive material, the refractive index increases with increasing frequency.

We choose a direction $\mathbf{s}$ of propagation (the angle θ) such that the intersecting plane (orthogonal to $\mathbf{s}$) cuts the inner ellipsoid in the points P_1 and P_2 of the plane of the figure. P_1 and P_2 are determined by the intersection of the sphere having a radius $n_{H1}(2f_1)$ with the inner spheroid. The ordinary wave at f_3 is polarized along the 1-direction, the extraordinary wave at f_1 in direction $\overline{P_1 P_2}$. Then, the ordinary wave and extraordinary wave possess the same index of refraction:

$$n_{(1)}(2f_1) = n_{H1}(2f_1) = n_{(2)}(f_1) \qquad (2.42)$$

The angle θ is determined analogously to Eq. (2.41) as

$$\tan \theta = \frac{n_{H3}(f_1)}{n_{H1}(f_1)} \sqrt{\frac{n_{H1}^2(f_3) - n_{H1}^2(f_1)}{n_{H3}^2(f_1) - n_{H1}^2(f_3)}} \qquad (2.43)$$

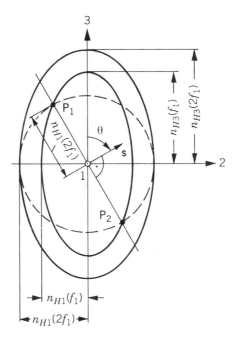

FIGURE 2.2   Phase matching.

With positively uniaxial media, phase matching is only possible if $n_{H3}(f_1) \geq n_{H1}(2f_1)$. Because of the rotational symmetry of the figure, the sphere intersects the inner spheroid in a circle. The position of **s** is only fixed by the angle θ to the 3-axis; the azimuthal angle is still free. Phase matching for negatively uniaxial media proceeds quite similarly.

Equality of refractive indices leads to equality of the phase velocities $v = c/n$:

$$v(f_3) \equiv v(2f_1) \stackrel{!}{=} v(f_1) \tag{2.44}$$

For

$$|\mathbf{k}| = k = k_0 n = \frac{\omega n(f)}{c} \qquad \mathbf{p} = \hbar \mathbf{k} \tag{2.45}$$

(**p** is the momentum of the photon), we find

$$\mathbf{k}_3 = \mathbf{k}_1 + \mathbf{k}_1, \quad \mathbf{p}_3 = \mathbf{p}_1 + \mathbf{p}_1$$

$$\Delta \mathbf{k} := \mathbf{k}_3 - 2 \cdot \mathbf{k}_1 = \frac{2\omega_1}{c} [n(2f_1) - n(f_1)] \mathbf{s} = 0 \tag{2.46}$$

(*Note*: $\mathbf{k}_3$, $\mathbf{k}_1$, $\mathbf{p}_3$, and $\mathbf{p}_1$ are all parallel to **s**). This is the conservation of momentum.

Particularly favorable is the case of $90°$ phase matching. Here $\Delta k = 0$ can be reached even for the case in which the parallel lying k vectors at $2f_1$ and f_1 (both parallel to **s**) lie simultaneously normal to the optical axis of the crystal ($\theta = 90°$). This is possible only for $n_{H1}(2f_1) = n_{(2)}(f_1) = n_{H3}(f_1)$. Thus, not only the phase fronts are parallel, but also the Poynting vectors $\mathbf{S}(f_1)$ and $\mathbf{S}(f_3)$ are parallel and are directed along the direction of **k**. Then, laterally bounded rays do not diverge if there is double refraction: no walk-off.

As already discussed, a phase mismatch with $\Delta k \neq 0$ is allowed. This means that conservation of momentum for the photons alone must not be fulfilled. In this case, the nonlinear medium is participating in the balance of momenta. However, one has the most intense light at the double frequency $f_3 = 2f_1$ only if $v(f_3) = v(f_1)$. Generally, the better the conditions of Eq. (2.47) are satisfied, the larger the effective energy transfer between the different fields:

$$\mathbf{k}_1 + \mathbf{k}_2 = \mathbf{k}_3 \qquad \text{or} \qquad \mathbf{p}_1 + \mathbf{p}_2 = \mathbf{p}_3 \tag{2.47}$$

In this general case, one has to draw three index ellipsoids for the three frequencies f_1, f_2, f_3. In collinear phase matching, all three wave vectors are taken to be parallel (see Problem 2.11). In type I phase matching for positively uniaxial material, both $n(f_1)$ and $n(f_2)$ are extraordinary, whereas $n(f_3)$ is ordinary. In type II phase matching, either $n(f_1)$ or $n(f_2)$ is ordinary, whereas $n(f_3)$ is ordinary. For negatively uniaxial material, the roles of ordinary and extraordinary waves are reversed.

PROBLEMS

Section 2.1

2.1 Derive Manley–Rowe equations for the process $f_3 = f_1 - f_2$ analogously to Eq. (2.8).

2.2 Derive Manley–Rowe equations for the general four-wave mixing process $f_4 = f_1 + f_2 + f_3$ from energy conservation (and losslessness).

2.3 Derive the Helmholtz equation in paraxial approximation starting from Maxwell's equations. *Hint*: Assume div $\mathbf{E} = 0$ and use Eq. (2.19), where now $\bar{\mathbf{E}} = \bar{\mathbf{E}}(x, y, z, f)$. The paraxial approximation corresponds to Eq. (2.20).

2.4 Derive Eq. (2.26).

Section 2.2

2.5 Show that a real value for $\bar{D}_x : \bar{D}_y$ corresponds to a linearly polarized wave.

2.6 Show analytically that $\mathbf{D}_{(1)} \cdot \mathbf{D}_{(2)} = 0$. *Hint*: Starting from Eq. (2.34), decompose $\mathbf{E} = \mathbf{E}_\parallel + \mathbf{E}_\perp$ into components parallel and orthogonal to $\mathbf{s}$ and show that $\mathbf{E}_{(2)} \cdot \mathbf{D}_{(1)} = \mathbf{E}_{(1)} \cdot \mathbf{D}_{(2)}$.

2.7 Assume that Φ is the angle between the wave vector $\mathbf{k}$ and the Poynting vector $\mathbf{S}$ (or between $\mathbf{D}$ and $\mathbf{E}$) in quartz ($n_e = 1.553, n_o = 1.544$). Sketch the dependence of Φ (or $\tan \Phi$) as a function of the angle θ between $\mathbf{k}$ and the 3-axis. Where does the maximum of Φ lie?

2.8 Derive Eq. (2.40).

2.9 Sketch the phase-matching procedure for negatively uniaxial media analogously to Fig. 2.2 and modify Eqs. (2.40) through (2.43) if necessary.

2.10 Consider SHG in a KDP crystal (negatively uniaxial) using as a pump wave the output of a pulsed ruby laser with $\lambda = 0.694 \ \mu$m. Calculate the phase-matching angle according to Eq. (2.41) with

$$n_e(\lambda = 0.694 \mu\text{m}) = 1.466, \qquad n_e(\lambda = 0.347 \mu\text{m}) = 1.487$$

$$n_o(\lambda = 0.694 \mu\text{m}) = 1.506, \qquad n_o(\lambda = 0.347 \mu\text{m}) = 1.534$$

2.11 Is collinear phase matching possible in three-wave mixing $\omega_3 = \omega_1 + \omega_2$, $k_3 = k_1 + k_2$?

BIBLIOGRAPHY

M. Born and E. Wolf. *Principles of Optics*. Pergamon Press, Oxford, UK, 1964.

G. K. Grau. *Quantenelektronik*. Vieweg, Braunschweig, 1978.

H. A. Haus. *Waves and Fields in Optoelectronics*. Prentice-Hall, Englewood Cliffs, NJ, 1984.

P. Meystre and M. Sargent III. *Elements of Quantum Optics*. Springer-Verlag, Berlin, 1990.

B. E. A. Saleh and M. C. Teich. *Fundamentals of Photonics*. Wiley, New York, 1991.

M. Schubert, B. Wilhelmi. *Nonlinear Optics and Quantum Electronics*. Wiley, New York, 1986.

A. Yariv. *Quantum Electronics*. Wiley, New York, 1975.

F. Zernike and J. E. Midwinter. *Applied Nonlinear Optics*. Wiley, New York, 1973.

CHAPTER THREE

Pockels Effect and Related Phenomena

We are interested here in processes $f_3 = f_1 + f_2$, for which one of the frequencies in the susceptibility tensor $\chi_{ijk}(-f_3,f_1,f_2)$ vanishes:

- $f_3 = 0 \longrightarrow f_2 = -f_1$ leads to optical rectification.
- $f_1 = 0 \longrightarrow f_2 = f_3$ yields the linear electro-optic effect or Pockels effect (correspondingly for $f_2 = 0$).

In Eq. (1.37), it was shown that for lossless processes the susceptibility tensors are real. The tensors for the linear electro-optic effect and optical rectification are connected, that is, in the lossless case, one has Eq. (1.38):

$$\chi_{ijk}(-f_1,f_1,0) = \chi_{kji}(0,f_1,-f_1) \tag{3.1}$$

In Eq. (1.46), we have shown that $\chi_{ijk}(0,f_1,-f_1)$ is symmetric in the last two tensor indices. Then, $\chi_{ijk}(-f_1,f_1,0)$ is symmetric in the first two tensor indices.

The linear electro-optic or Pockels effect had been detected already in 1893, many years before the advent of strong laser light sources. This was feasible since it was here, as it was with the quadratic electro-optic or Kerr effect, a question of the influence of a strong homogeneous dc field $\hat{\mathbf{E}}(f = 0)$ (or of a low-frequency field) on the propagation properties of a medium for optical waves, whose amplitudes may be very small in principle. As a nonlinear effect of second order, Eq. (1.56), the Pockels effect occurs only in crystals without inversion symmetry; see Eq. (1.36). The dc field induces a double refraction, that is, the permittivity tensor will be changed and therefore the refractive index too. Of course, this change is small; however, the effect can produce a significant change at distances that are large compared with the wavelength.

In the Faraday effect, the electric dc field is replaced by a magnetic dc field.

3.1   THE LINEAR ELECTRO-OPTIC EFFECT

We first discuss the change of the permittivity tensor ε_{ij} if an additional electric dc field is switched on. Then the linear electro-optic effect in KDP is treated, which allows application as polarization or intensity modulator, light deflector, and so on.

3.1.1 Theoretical Description

If we introduce in Eq. (1.56):

$$\widehat{P}_i^{(2)}(f_1) = 2\varepsilon_0 \sum \chi_{ijk}(-f_1, f_1, 0)\widehat{E}_j(f_1)\widehat{E}_k(0)$$

the definition

$$\tilde{\chi}_{ij}[f_1, \widehat{E}(0)] := 2\sum_k \chi_{ijk}(-f_1, f_1, 0)\widehat{E}_k(0) = 2\sum_k \chi_{kji}(0, f_1, -f_1)\widehat{E}_k(0)$$

where we have used Eq. (3.1), we obtain the total polarization as the sum of the linear and the nonlinear polarization:

$$
\begin{aligned}
\widehat{P}_i(f_1) &= \widehat{P}_i^{(1)}(f_1) + \widehat{P}_i^{(2)}(f_1) \\
&= \varepsilon_0 \sum \chi_{ij}(f_1)\widehat{E}_j(f_1) + \varepsilon_0 \sum \tilde{\chi}_{ij}[f_1, \widehat{E}(0)]\widehat{E}_j(f_1) \\
&= \varepsilon_0 \sum \chi_{ij}^{\text{tot}}[f_1, \widehat{E}(0)]\widehat{E}_j(f_1)
\end{aligned}
\tag{3.2}
$$

where

$$\chi_{ij}^{\text{tot}}[f_1, \widehat{E}(0)] = \chi_{ij}(f_1) + \tilde{\chi}_{ij}(f_1, \widehat{E}(0)) \tag{3.3}$$

Thus, the relation

$$\varepsilon_{ij}(f_1) = \varepsilon_0[\delta_{ij} + \chi_{ij}(f)]$$

valid in the case of a vanishing field $\widehat{E}(0)$ (linear case), is transformed into

$$\varepsilon_{ij}^{\text{tot}}[f_1, \widehat{E}(0)] = \varepsilon_0[\delta_{ij} + \chi_{ij}(f_1) + \tilde{\chi}_{ij}] = \varepsilon_{ij}(f_1) + \varepsilon_0\tilde{\chi}_{ij}[f_1, \widehat{E}(0)] \tag{3.4}$$

if the additional dc field $\widehat{E}(0)$ is switched on. The permittivity $\varepsilon_{ij}^{\text{tot}}[f_1, \widehat{E}(0)]$ now depends on the dc field $\widehat{E}(0)$, as does the refractive index.

Without the dc field, the index ellipsoid in the system of the principal dielectric axes has the form (optically uniaxial medium, $n_{H1} = n_{H2} = n_o$):

$$1 = \sum_{i=1}^{3} \frac{x_i^2}{n_{Hi}^2} = (x_1, x_2, x_3) \cdot \begin{bmatrix} \frac{1}{n_{H1}^2} & 0 & 0 \\ 0 & \frac{1}{n_{H1}^2} & 0 \\ 0 & 0 & \frac{1}{n_{H3}^2} \end{bmatrix} \cdot \begin{bmatrix} x_1 \\ x_2 \\ x_3 \end{bmatrix} = \mathbf{x}^T \mathbf{M}_0 \mathbf{x} \qquad (3.5)$$

with

$$(M_0)_{ij} = \frac{\delta_{ij}}{n_{Hi}^2} = \frac{\delta_{ij}}{(\varepsilon_{Hi}/\varepsilon_0)} = \frac{\delta_{ij}}{(\varepsilon_{ij}/\varepsilon_0)}$$

Thus, $\mathbf{M}_0$ is the inverse tensor to $\varepsilon_{ij}/\varepsilon_0$. $\mathbf{x}$ represents a column matrix: $\mathbf{x}^T = (x_1, x_2, x_3)$.

However, with the electrical dc field switched on, the index ellipsoid will change and takes the general form

$$\mathbf{x}^T \cdot \mathbf{M} \cdot \mathbf{x} = \sum M_{ij} x_i x_j = 1 \qquad (3.6)$$

We use the ansatz

$$M_{ij} = (M_0)_{ij} + (\Delta M_0)_{ij} = \frac{\delta_{ij}}{n_{Hi}^2} + \sum_k r_{ijk} \widehat{E}_k(0) \qquad (3.7)$$

and M_{ij} can be calculated again from

$$\delta_{ik} = \sum_j M_{ij} \cdot \left(\frac{\varepsilon_{jk}^{tot}}{\varepsilon_0} \right)$$

as an inverse tensor. If we substitute here Eqs. (3.4) and (3.7) and neglect terms that are quadratic in the electric field, we get [*Note*: in the system of principal axes belonging to the case $\widehat{E}(0) = 0$, the tensor ε_{ij} but not ε_{ij}^{tot}, in Eq. (3.4) is diagonal],

$$r_{ijk} = -\frac{2\chi_{kij}(0, f_1, -f_1)}{n_{Hi}^2 \cdot n_{Hj}^2} \qquad (3.8)$$

$\chi_{kij}(0, f_1, -f_1)$ is symmetric in the two last indices (ij), and so r_{ijk} is symmetric in the first two indices. Therefore, we introduce a 6×3 matrix with the definition

$(k = 1, 2, 3)$:

$$r_{11k} = r_{1k}, \qquad r_{22k} = r_{2k}, \qquad r_{33k} = r_{3k},$$

$$r_{23k} = r_{32k} = r_{4k}, \qquad r_{13k} = r_{31k} = r_{5k}, \qquad r_{12k} = r_{21k} = r_{6k} \tag{3.9}$$

Then, the index ellipsoid, Eq. (3.6), is, with Eq. (3.7),

$$1 = \sum_{i=1}^{3} \frac{x_i^2}{n_{Hi}^2} + (x_1^2 \; x_2^2 \; x_3^2 \; 2x_2x_3 \; 2x_1x_3 \; 2x_1x_2) \cdot \begin{bmatrix} r_{11} & r_{12} & r_{13} \\ r_{21} & r_{22} & r_{23} \\ r_{31} & r_{32} & r_{33} \\ r_{41} & r_{42} & r_{43} \\ r_{51} & r_{52} & r_{53} \\ r_{61} & r_{62} & r_{63} \end{bmatrix} \cdot \begin{bmatrix} \hat{E}_1(0) \\ \hat{E}_2(0) \\ \hat{E}_3(0) \end{bmatrix} \tag{3.10}$$

The 6×3 scheme of the electro-optic coefficients r_{ij} is often called the *electro-optic tensor*. Symmetry considerations show which one of the 18 elements of (r_{ij}) vanishes and which relations exist between the remaining elements. Electro-optic coefficients have an order of magnitude of about 10^{-10} to 10^{-12} m/V.

The principal axes of the new index ellipsoid can again be found by a diagonalization of the symmetric real (3×3) matrix **M**.

3.1.2 Application

Let us consider the electro-optic effect in KDP (KH_2PO_4) or ADP ($NH_4H_2PO_4$). From the symmetry group ($\overline{4}$2m) of these two negatively uniaxial crystals, it results that all elements of the electro-optic tensor vanish except $r_{41} = r_{52}$ and r_{63}, that is, the index ellipsoid is reduced to the form

$$\frac{x_1^2}{n_{H1}^2} + \frac{x_2^2}{n_{H1}^2} + \frac{x_3^2}{n_{H3}^2} + 2r_{41}\hat{E}_1(0)x_2x_3 + 2r_{41}\hat{E}_2(0)x_1x_3 + 2r_{63}\hat{E}_3(0)x_1x_2 = 1 \tag{3.11}$$

We assume that the applied dc field points in the 3-direction: $\hat{\mathbf{E}}(0) = \hat{E}_3(0)\mathbf{e}_3$, the direction of the optical axis, and we now get for the index ellipsoid

$$\frac{x_1^2}{n_{H1}^2} + \frac{x_2^2}{n_{H1}^2} + \frac{x_3^2}{n_{H3}^2} + 2r_{63}\hat{E}_3(0)x_1x_2 = 1 \tag{3.12}$$

Thus, because of the special choice of the dc field and the symmetry of the crystal, a rotation in the x_1, x_2 plane by an angle of $45°$ is sufficient for diagonalization:

$$x_1 = x_1' \cos 45° - x_2' \sin 45° = \frac{(x_1' - x_2')}{\sqrt{2}}$$

$$x_2 = x_1' \sin 45° + x_2' \cos 45° = \frac{(x_1' + x_2')}{\sqrt{2}} \qquad (3.13)$$

Substitution in Eq. (3.12) yields

$$\left[\frac{1}{n_{H1}^2} + r_{63}\widehat{E}_3(0) \right] x_1'^2 + \left[\frac{1}{n_{H1}^2} - r_{63}\widehat{E}_3(0) \right] x_2'^2 + \frac{x_3^2}{n_{H3}^2} = 1 \qquad (3.14)$$

or

$$\frac{x_1'^2}{n_1'^2} + \frac{x_2'^2}{n_2'^2} + \frac{x_3^2}{n_{H3}^2} = 1 \qquad (3.15)$$

The new principal axes appear rotated in the x_1, x_2 plane by $+45°$; the lengths of the principal axes are given by n_{H3} (unchanged) and

$$\frac{1}{n_1'^2}_{2} = \frac{1}{n_{H1}^2} \pm r_{63}\widehat{E}_3(0)$$

or

$$n_1' = n_{x'} = \left[\frac{1}{n_{H1}^2} \pm r_{63}\widehat{E}_3(0) \right]^{-1/2} \approx \sqrt{n_{H1}^2 [1 \mp n_{H1}^2 r_{63}\widehat{E}_3(0)]}$$

$$\approx n_{H1} \cdot \left[1 \mp \frac{n_{H1}^2 r_{63}\widehat{E}_3(0)}{2} \right] = n_{H1} \mp \frac{n_{H1}^3 r_{63}\widehat{E}_3(0)}{2} = n_o \mp \frac{n_o^3 r_{63}\widehat{E}_3(0)}{2} \qquad (3.16)$$

They are thus dependent on $\widehat{E}_3(0)$, and the medium is now optically biaxial.

Example 3.1   Typical values of r_{ij} lie in the range 10^{-12} to 10^{-10} m/V. If one applies a voltage of 10 kV across a Pockels cell of thickness 1 cm, the term $\frac{1}{2} n_o^3 r_{63}\widehat{E}_3(0)$ is of the order of 10^{-6} to 10^{-4}.   ∎

Figure 3.1 shows two Pockels cells: (1) in the longitudinal mode of operation and (2) in the transversal mode of operation, where the applied field is $\widehat{\mathbf{E}}(0) = \widehat{E}_1(0)\mathbf{e}_1$. (See Problem 3.4).

How does a wave incident in the 3-direction and linearly polarized along the former y-direction propagate? It can be decomposed in the plane $z = 0$ into two orthogonal linearly polarized waves in the x'- and y'-direction. After propaga-

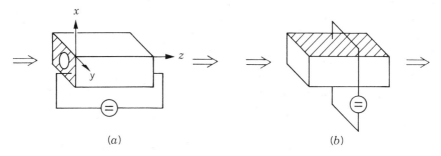

FIGURE 3.1   Electro-optic modulators: (*a*) longitudinal mode and (*b*) transverse mode.

tion, one obtains at the position z for the components of $\widehat{\mathbf{E}}(\mathbf{x}, f)$

$$\widehat{E}_{x'} = A \, \exp\left(-j\frac{\omega}{c}n_{x'}z\right) = A \, \exp\left\{-j\frac{\omega}{c}z\left[n_o - n_o^3 r_{63}\frac{\widehat{E}_3(0)}{2}\right]\right\}$$

$$\widehat{E}_{y'} = A \, \exp\left(-j\frac{\omega}{c}n_{y'}z\right) = A \, \exp\left\{-j\frac{\omega}{c}z\left[n_o + n_o^3 r_{63}\frac{\widehat{E}_3(0)}{2}\right]\right\}$$

(3.17)

or abbreviated

$$\widehat{E}_{x'} = A \cdot e^{-j(\psi - \phi)}, \qquad \widehat{E}_{y'} = A \cdot e^{-j(\psi + \phi)}$$

In the plane $z = L$, a phase difference (electro-optic retardation) exists between both components:

$$\Gamma := 2\phi = \left(\frac{\omega}{c}\right) L n_o^3 r_{63} \widehat{E}_3(0) = \left(\frac{\omega}{c}\right) n_o^3 r_{63} U = \pi \cdot \frac{U}{U_\pi} \qquad (3.18)$$

with the dc voltage $U = \widehat{E}_3(0)L$. $U_\pi = \lambda/(2n_o^3 r_{63})$ is the so-called *half-wave voltage*: for $U = U_\pi$, one finds $\Gamma = \pi$. (Numerical value for ADP: $U_\pi = 10 \, \mathrm{kV}$ at $\lambda = 0.5 \, \mu\mathrm{m}$, since $n_o^3 r_{63} = 27 \times 10^{-12}$ m/V.)

At $z = 0$ or for $U = 0$, one has $\Gamma = 0$: In the new system x', y', the wave is linearly polarized along the $+45°$ direction; this is the original y-axis. If we choose $U = U_\pi/2$, or $\Gamma = \pi/2$, we get, omitting a common phase factor,

$$E_{x'} = \mathrm{Re} \, \widehat{E}_{x'} \exp j\omega t = A \cdot \cos \omega t$$

$$E_{y'} = \mathrm{Re} \, \widehat{E}_{y'} \exp j\omega t = A \cdot \cos\left(\omega t - \frac{\pi}{2}\right) = A \cdot \sin \omega t$$

(3.19)

and we find a circular polarization. If finally $U = U_\pi$ or $\Gamma = \pi$, we have

$$E_{x'} = \mathrm{Re} \, \widehat{E}_{x'} \exp j\omega t = A \cdot \cos \omega t$$

$$E_{y'} = \mathrm{Re} \, \widehat{E}_{y'} \exp j\omega t = A \cdot \cos(\omega t - \pi) = -A \cdot \cos \omega t$$

(3.20)

One gets in the new system x', y', a wave linearly polarized along the $-45°$ direction (in the direction of the original x-axis); thus, the polarization is rotated $90°$, compared with the incident polarization.

Instead of the dc field, one can also use a field with low frequency, compared to optical frequencies (f_{mod} up to 1 GHz). The effect can then be applied for polarization modulation, or together with an analyzer as an intensity modulator, or for light deflection, and so on.

Example 3.2

How can the Pockels effect be used for intensity modulation?

The incident field is parallel to the y-direction; therefore, we take $\bar{E}_{x'}(z = 0) = \bar{E}_{y'}(z = 0) = A = \text{const}$ with an input intensity

$$I_y(0) = |\bar{E}_{x'}(0) \sin 45° + \bar{E}_{y'}(0) \cos 45°|^2 = 2A^2.$$

Then $\bar{E}_{x'}(L) = A, \bar{E}_{y'}(L) = A \exp(-j\Gamma)$ (up to a common phase term). Behind an output polarizer (in the x-direction), one finds at $z = L$ a total field of

$$\bar{E}_x(L) = [\bar{E}_{x'}(L) - \bar{E}_{y'}(L)] \cos 45° = A[1 - \exp(-j\Gamma)]/\sqrt{2}$$

Therefore, as transmittance,

$$T = \frac{I(L)}{I(0)} = \frac{2A^2 \sin^2 \Gamma/2}{2A^2} = \sin^2 \frac{\Gamma}{2} = \sin^2\left(\frac{\pi}{2}\frac{U}{U_\pi}\right)$$

Now, the modulator is biased by choosing a slowly varying voltage

$$U = \frac{U_\pi}{2} + U_m \sin \omega_m t$$

with the modulation frequency ω_m and a modulation voltage U_m. Then, we get for the transmittance

$$T = \sin^2\left[\frac{\pi}{2U_\pi}\left(\frac{U_\pi}{2} + U_m \sin \omega_m t\right)\right] = \sin^2 \frac{1}{2}\left(\frac{\pi}{2} + \Gamma_m \sin \omega_m t\right)$$

$$= \frac{1}{2}[1 + \sin(\Gamma_m \sin \omega_m t)] \approx \frac{1}{2}(1 + \Gamma_m \sin \omega_m t)$$

with $\Gamma_m = \pi(U_m/U_\pi)$ and a fixed retardation of $\pi/2$. The approximation is valid for $\Gamma_m \ll 1$. We see that T is now approximately a linear function of the modulation voltage U_m. ∎

3.2   OPTICAL RECTIFICATION

As already mentioned in Eq. (3.1), the susceptibility tensor $\chi_{ijk}(0,f_1,-f_1)$ for optical rectification is connected to the tensor of the linear electro-optic effect $\chi_{ijk}(-f_1,f_1,0)$.

In optical rectification, light from an intense polarized optical field is sent through a dielectric slab made of an appropriate nonlinear material (optically uniaxial). In the slab, a constant static polarization is generated. It looks as if the slab had been brought into the static electric field between the plates of a capacitor; a dc voltage will be induced.

As an example, we consider an uncharged capacitor, with its plates parallel to the xy-plane. Between these plates is a slab of KDP; its optical axis points in the z-direction, normal to the condenser plates. The KDP material is completely filled by the optical wave and does not touch the condenser plates. The propagation vector $\mathbf{k} = k\mathbf{s} = (\omega n/c)\mathbf{s}$ lies in the xy-plane. For the optical $\mathbf{D}$ field ($\mathbf{e}$ is a unit vector in the direction of $\widehat{\mathbf{D}}$), we set

$$\widehat{\mathbf{D}} = A \cdot \exp(-jk\mathbf{s} \cdot \mathbf{x}) \cdot \mathbf{e}, \qquad \widehat{\mathbf{D}} \| \mathbf{e} \perp \mathbf{s} \tag{3.21}$$

The incident electric field $\mathbf{D}$ then propagates as an ordinary wave with a $\mathbf{D}$ vector ($\| \mathbf{E}$ vector), which lies in the xy-plane ($\perp \mathbf{s}$): $\mathbf{E}_o = E_x\mathbf{e}_x + E_y\mathbf{e}_y$, and as an extraordinary wave with a $\mathbf{D}$ vector ($\| \mathbf{E}$ vector) in the direction of the z-axis: $\mathbf{E}_e = E_z\mathbf{e}_z$. The susceptibility tensor $\chi_{ijk}(0,f_1,-f_1)$ of KDP vanishes, if not all indices are different (numerical values at $\lambda = 1.06\,\mu\text{m}$: $\chi_{123} = \chi_{231} = \chi_{132} = \chi_{213} = 1.01 \times 10^{-11}\,\text{cm/V}$, $\chi_{312} = \chi_{321} = 1.00 \times 10^{-11}\,\text{cm/V}$). Therefore, the extraordinary part does not yield a dc polarization since both E fields in Eq. (1.44):

$$\widehat{P}_i(0) = \left(\frac{\varepsilon_0}{2}\right) \sum \chi_{ijk}(0,f_1,-f_1)\widehat{E}_j(f_1)\widehat{E}_k^*(f_1) \tag{3.22}$$

possess in this case only a z-component: $j = k = 3$. However, the ordinary part yields a polarization $\widehat{\mathbf{P}}(0) = \widehat{P}_z(0)\mathbf{e}_z$, where $\widehat{P}_z(f = 0)$ is independent of $\mathbf{x}$, because the spatially dependent exponents in Eq. (2.19) cancel if inserted into Eq. (3.22). The electric field $\widehat{E}_{\text{ind}}$ induced by this polarization is determined—in the case of no charges—by the electrostatic equations

$$\nabla \times \widehat{\mathbf{E}}_{\text{ind}} = 0, \qquad \nabla \cdot \widehat{\mathbf{D}} = \nabla \cdot [\varepsilon_0\widehat{\mathbf{E}}_{\text{ind}} + \widehat{\mathbf{P}}(0)] = 0$$

This is fulfilled by $\widehat{\mathbf{E}}_{\text{ind}} = (1/c_0)\widehat{\mathbf{P}}(0)$. With an optical pulse of several MW peak power, one obtains in KDP a voltage of several $100\,\mu\text{V}$. However, this effect has not found any practical application.

3.3 MAGNETO-OPTIC EFFECTS (FARADAY EFFECT)

An originally optically isotropic medium can become optically biaxial under the influence of an external magnetic dc field **H** and can show double refraction. Thus, the state of polarization can be changed. Via a classical model, we derive now the corresponding permittivity tensor.

The displacement **x** of electrons from the equilibrium position will be described approximately by an undamped forced oscillation:

$$m\ddot{\mathbf{x}} + K \cdot \mathbf{x} = -e\mathbf{E} - e(\dot{\mathbf{x}} \times \mathbf{B}) \tag{3.23}$$

with $\mathbf{B} = \mu_0 \mathbf{H}$. The ansatz $\mathbf{x} = \mathbf{x}_0 \cdot \exp(j\omega t)$ yields

$$(-\omega^2 m + K) \cdot \mathbf{x} = -e\mathbf{E} - ej\omega(\mathbf{x} \times \mathbf{B})$$

The polarization is again given by $\mathbf{P} = -Ne\mathbf{x}$, where N is the number of charge carriers (dipoles) per unit volume (see Section 1.4). After multiplication with $-Ne$, one finds

$$\alpha \cdot \mathbf{P} = +Ne^2\mathbf{E} - ej\omega(\mathbf{P} \times \mathbf{B})$$

with the abbreviation $\alpha = K - \omega^2 m$. If the magnetic field has only a z-component: $\mathbf{B} = B \cdot \mathbf{e}_z$, one obtains the following system of equations:

$$
\begin{array}{ccccccc}
\alpha \cdot P_x & + & j\omega e B P_y & + & 0 \cdot P_z & = & Ne^2 E_x \\
-j\omega e B P_x & + & \alpha \cdot P_y & + & 0 \cdot P_z & = & Ne^2 E_y \\
0 \cdot P_x & + & 0 \cdot P_y & + & \alpha \cdot P_z & = & Ne^2 E_z
\end{array}
$$

or $P_z = (Ne^2/\alpha) \cdot E_z$ and

$$
\begin{bmatrix} P_x \\ P_y \end{bmatrix} = \frac{Ne^2}{\alpha^2 - e^2\omega^2 B^2}
\begin{bmatrix} \alpha & -j\omega e B \\ j\omega e B & \alpha \end{bmatrix}
\begin{bmatrix} E_x \\ E_y \end{bmatrix} \tag{3.24}
$$

Comparing with the expression $\mathbf{P} = \varepsilon_0 \cdot \mathbf{\chi} \cdot \mathbf{E}$, we see that the susceptibility tensor $\chi(B)$ has the form

$$
(\chi_{ij}) =
\begin{bmatrix}
\chi_{xx}(B^2) & \chi_{xy}(B) & 0 \\
\chi_{yx}(B) & \chi_{xx}(B^2) & 0 \\
0 & 0 & \chi_{zz}
\end{bmatrix}
$$

$$
=
\begin{bmatrix}
\chi'_{xx} & 0 & 0 \\
0 & \chi'_{xx} & 0 \\
0 & 0 & \chi'_{zz}
\end{bmatrix}
+ j
\begin{bmatrix}
0 & -\chi''_{yx} & 0 \\
\chi''_{yx} & 0 & 0 \\
0 & 0 & 0
\end{bmatrix} \tag{3.25}
$$

Thus, the components $\chi_{ij} = \chi'_{ij} + j \cdot \chi''_{ij}$ depend on B, with the exception of χ_{zz}. One finds immediately the following properties:

$$\chi_{ij}(B) = \chi^*_{ji}(B) = \chi_{ji}(-B)$$

$$\chi'_{ij}(B) = \chi'_{ji}(B) = \chi'_{ji}(-B), \qquad \chi''_{ij}(B) = -\chi''_{ji}(B) = \chi''_{ji}(-B) \tag{3.26}$$

Therefore, the real part χ'_{ij} is symmetric in the tensor indices and an even function of B; it produces a linear double refraction (Cotton–Mouton effect); see Problem 3.7. The imaginary part χ''_{ij} is antisymmetric and an odd function of B, it produces a circular double refraction (Faraday effect). From these two magneto-optical effects, we shall only discuss the Faraday effect (terms $\sim B$ are required), since the Cotton–Mouton effect is much weaker (terms $\sim B^2$) and therefore has not found important application. In contrast, the Faraday effect is used, for example, for an optical isolator that prevents, for instance, reflected light from reentering a laser cavity (see Fig. 3.2 and Problem 3.5). We neglect then terms of higher order as B and obtain

$$\chi_{xx} = \chi_{yy} = \chi_{zz} = \chi'_{xx} = \chi'_{yy} = \chi'_{zz} = \frac{Ne^2}{(\varepsilon_0 \alpha)}$$

$$\chi_{yx} = 0 + j\chi''_{yx} = \frac{j\omega Ne^3 B}{(\varepsilon_0 \alpha^2)} \tag{3.27}$$

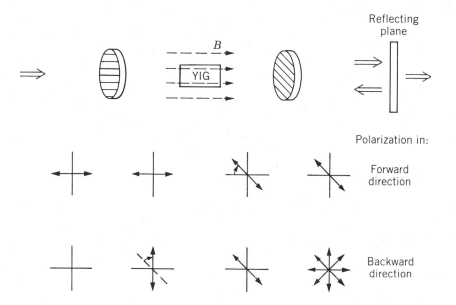

FIGURE 3.2 Optical isolator using Faraday effect.

Quite generally, we define formally a third rank tensor χ_{ijk}'' by writing

$$\chi_{ij}''(B) =: \chi_{ijz}'' \cdot B = \sum_k \chi_{ijk}'' \cdot B_k$$

$(B_z = B, B_x = B_y = 0)$; therefore,

$$P_i = \varepsilon_0 \sum_j (\chi_{ij}' + j \cdot \chi_{ij}'') \cdot E_j = \varepsilon_0 \sum_j \chi_{ij}' E_j + j\varepsilon_0 \sum_{jk} \chi_{ijk}'' E_j B_k \qquad (3.28)$$

The Faraday effect is thus a phenomenon with second-order polarization, in which an electric field and a magnetic field interact via the imaginary part of a "susceptibility tensor." Equation (3.28) corresponds to Eq. (3.2) for the Pockels effect, if one replaces the constant magnetic flux density B by $E(0)$.

According to Eq. (1.21), the dielectric tensor $\varepsilon_{ij}/\varepsilon_0$ has the same structure as the susceptibility tensor, Eq. (3.25):

$$\frac{(\varepsilon_{ij})}{\varepsilon_0} = \begin{bmatrix} 1+\chi_{xx} & -j\chi_{yx}'' & 0 \\ j\chi_{yx}'' & 1+\chi_{xx} & 0 \\ 0 & 0 & 1+\chi_{xx} \end{bmatrix} = \frac{1}{\varepsilon_0}\begin{bmatrix} \varepsilon_{xx} & -j\varepsilon_{yx}'' & 0 \\ j\varepsilon_{yx}'' & \varepsilon_{xx} & 0 \\ 0 & 0 & \varepsilon_{xx} \end{bmatrix} \qquad (3.29)$$

We look now for the new principal dielectric axes, where $\mathbf{D} \parallel \mathbf{E}$, or

$$\sum_j \varepsilon_{ij} E_j^{(e)} = \lambda E_i^{(e)}$$

A diagonalization of Eq. (3.29) yields the eigenvalues λ and the normalized and constant eigenvectors $\mathbf{E}^{(e)}$ (biaxial medium!)

$$\lambda = \varepsilon_{xx} \pm \varepsilon_{yx}'' =: \varepsilon_\pm, \qquad \mathbf{E}_\pm^{(e)} = \frac{1}{\sqrt{2}} \cdot (\mathbf{e}_x \pm j \cdot \mathbf{e}_y) = \mathbf{e}_\pm \qquad (3.30)$$

[and $\lambda = \varepsilon_{xx}$ with $\mathbf{E}_3^{(e)} = \mathbf{e}_z$]. These polarization unit vectors $\mathbf{e}_\pm$ represent right and left circularly polarized light with the propagation constants

$$k_\pm = \frac{\omega\sqrt{\varepsilon_\pm/\varepsilon_0}}{c} \approx \frac{\omega\sqrt{\varepsilon_{xx}/\varepsilon_0}}{c}\left(1 \pm \frac{\varepsilon_{yx}''}{2\varepsilon_{xx}}\right) \qquad (3.31)$$

An arbitrary circularly polarized wave then has the form

$$\tilde{\mathbf{E}}_\pm = \frac{E_0}{\sqrt{2}} \exp[j(\omega t - k_\pm z)] \cdot \mathbf{e}_\pm$$

$$= \frac{E_0}{\sqrt{2}} \exp\left[j(\omega t - \frac{k_+ + k_-}{2}\cdot z) \mp j\frac{k_+ - k_-}{2}\cdot z\right]\cdot \mathbf{e}_\pm$$

$$= \frac{E_0}{\sqrt{2}} \exp(j\phi \mp j\delta) \cdot \mathbf{e}_\pm \qquad (3.32)$$

A wave incident at $z = 0$ linearly polarized along the x-direction is decomposed as

$$\widetilde{\mathbf{E}}(z = 0) = E_0 \exp(j\omega t)\mathbf{e}_x = \frac{E_0}{\sqrt{2}} \cdot \exp(j\omega t)(\mathbf{e}_+ + \mathbf{e}_-)$$

$$= \widetilde{\mathbf{E}}_+(z = 0) + \widetilde{\mathbf{E}}_-(z = 0) \qquad (3.33)$$

Then, we get for $z > 0$

$$\widetilde{\mathbf{E}}(z) = \widetilde{\mathbf{E}}_+(z) + \widetilde{\mathbf{E}}_-(z) = \frac{E_0}{\sqrt{2}} \cdot [\exp j(\phi - \delta) \cdot \mathbf{e}_+ + \exp j(\phi + \delta) \cdot \mathbf{e}_-]$$

$$= E_0 \exp(j\phi)(\cos \delta \cdot \mathbf{e}_x + \sin \delta \cdot \mathbf{e}_y)$$

or

$$\tilde{E}_x(z) = E_0 \exp(j\phi) \cdot \cos(\delta)$$
$$\tilde{E}_y(z) = E_0 \exp(j\phi) \cdot \sin(\delta) \qquad (3.34)$$

This means again linearly polarized light combined with a rotation of the plane of vibration from $\delta = 0$ at $z = 0$ to the value

$$\delta(L) = \frac{k_+ - k_-}{2} \cdot L \approx \frac{\omega}{2c} \cdot \frac{\varepsilon''_{yx}}{\sqrt{\varepsilon_0 \varepsilon_{xx}}} \cdot L \qquad (3.35)$$

at $z = L$. But according to Eqs. (3.29) and (3.27), ε''_{yx} is proportional to H for small H; thus,

$$\delta = V \cdot H \cdot L \qquad (3.36)$$

V is the so-called *Verdet constant*. For quartz, $V = 0.0209$ min/A at $\lambda = 589$ nm. The Verdet 'constants' depend on the wavelength λ; for example, extra-heavy lead glass from Schott (SFS-6) has $V = 0.214$ min/A ($\lambda = 0.5\ \mu$m), $V = 0.0402$ min/A ($\lambda = 1.0\ \mu$m), $V = 0.0176$ min/A ($\lambda = 1.5\ \mu$m). Very often, paramagnetic glasses doped with rare earths are used. In the near infrared ($\lambda = 1.2$ and $1.5\ \mu$m), one takes a YIG crystal (yttrium-iron-garnet = $Y_3Fe_5O_{12}$) with $V \approx 10$ min/A.

One obtains the Verdet constant in esu units [min/(cm · Oe)] by multiplication of the numerical values above with $10/(4\pi)$.

Example 3.3   Carbon disulfide has a Verdet constant $V - 0.0529$ min/A. Show that for $L = 1$ cm and $H = 796,000$ A/m, we get a rotation of the plane of polarization of $\delta = 7°1.1'$.

PROBLEMS

Section 3.2

3.1 Calculate the retardation Γ for an ADP crystal with $E_3(0) = 10^6 \text{V/m}$ and $L = 1$ cm at $\lambda = 0.5 \mu\text{m}$.

3.2 Show that all elements of the electro-optic tensor r_{ij} for a crystal of the symmetry group $(\bar{4}2m)$ vanish, except $r_{41} = r_{52}$ and r_{63}.

3.3 Repeat the calculation in Section 3.1.2 for GaAs ($n_{H1} = n_{H2} = n_{H3}$); where $r_{41} = r_{52} = r_{63}$, all the other elements vanish. Calculate $n_1'(E), n_2'(E)$, and $n_3'(E)$ for an electric field pointing in the 3-direction.

3.4 In Section 3.1.2, the electric field was applied along the direction of light propagation (the z-direction). This is the so-called longitudinal mode of modulation; see Fig. 3.1.

    (a) Discuss the disadvantages of these devices.

    (b) In the transverse mode, the field is applied in a direction perpendicular to that of light propagation. Calculate the new principal refractive indices and the retardation for GaAs in this mode.

Section 3.4

3.5 Light is incident from the left on two polarizers P and P$'$, which are separated by a column of carbon disulfide CS$_2$, 0.5 m long; the Verdet constant $V = 0.0529$ min/A. The plane of polarization of P$'$ is rotated by $+45°$ with respect to P.

    (a) Calculate the magnetic field **H** necessary for light transmitting P$'$.

    (b) What happens if all remains unchanged, but the light is transmitted from right to left? Explain Fig. 3.2.

3.6 A beam of linearly polarized light is sent through a piece of solid glass tubing 25 cm long and 1 cm in diameter. The tubing is wound with a single layer of 250 turns of copper wire along its entire length (long thin coil). If the Verdet constant of the glass is 0.06 min/A, what is the amount of rotation of the plane of polarization of the light when a current of 1.5 A is flowing through the wire?

3.7 Show that in the Cotton–Mouton effect, one has two linearly polarized, orthogonal waves, or linear birefringence. *Hint*: Take now $\mathbf{k} = k\mathbf{e}_x, D_x = 0$.

BIBLIOGRAPHY

R. W. Boyd. *Nonlinear Optics*. Academic Press, San Diego, CA, 1992.

P. N. Butcher. *Nonlinear Optical Phenomena*. Bulletin 200, Engineering Experimental Station, Ohio State University, Columbus, Ohio, USA.

G. R. Fowler. *Introduction to Modern Optics*. 2nd ed. Holt Rinehart and Winston, New York, 1975.

G. K. Grau. *Quantenelektronik*. Vieweg, Braunschweig, 1978.

K. D. Möller. *Optics*. University Science Books, Mills Valley, CA, 1988.

A. Nussbaum and R. A. Phillips. *Contemporary Optics*. Prentice-Hall, Englewood Cliffs, NJ, 1976.

B. E. A. Saleh and M. C. Teich. *Fundamentals of Photonics*. Wiley, New York, 1991.

Y. R. Shen. *The Principles of Nonlinear Optics*. Wiley, New York, 1984.

A. Yariv. *Quantum Electronics*. Wiley, New York, 1975.

CHAPTER FOUR

Second Harmonic Generation

Second harmonic generation belongs to the most important nonlinear processes since it is relatively easy to realize and yields a relatively high efficiency. We discuss this process together with some supplements.

4.1 QUALITATIVE TREATMENT

Second harmonic generation (SHG) [Eq. (1.43)] is a process with quadratic polarization where only one single monochromatic field with frequency f_1 is incident. The field $\mathbf{E}_0(f_1)$ (red light, for example) propagates with a phase velocity $v_1 = c/n(f_1)$ in the nonlinear crystal and generates the polarization wave $\mathbf{P}(2f_1)$. The point where this wave is generated propagates, of course, just with the same phase velocity v_1. However, this polarization induces in the nonlinear medium an electric field $\mathbf{E}_d(2f_1)$ (blue light). This second harmonic propagates now with the phase velocity $v_2 = c/n(2f_1)$. Therefore, the second harmonic wave $\mathbf{E}_d(2f_1)$ and the fundamental $\mathbf{E}_0(f_1)$ fall out of step with the increasing distance in the crystal, and thus, the second harmonic waves generated at different places of the crystal do no longer superpose in phase; an extinction of the second harmonic wave by destructive interference results. Subsequently, the same process can start anew. Therefore, the intensity of the second harmonic will spatially decrease and increase.

Intense light in crystals can be obtained by phase matching with $v(f_1) = v(2f_1)$ or $n(f_1) = n(2f_1)$; see Eq. (2.44). Then, the polarization wave and the second harmonic remain in phase; one has a superposition with a constant phase relation and an increase of amplitude in the second harmonic with increasing thickness of the crystal ($\sim L$), so long as the total intensity of the fundamental wave has been totally converted. Thus, the intensity must increase $\sim L^2$.

In a corpuscular (quantum theoretical) picture, two photons of energy hf_1 are annihilated and simultaneously a photon of energy $h \cdot 2f_1$ is created. Only half as many photons are generated as annihilated.

4.2   QUANTITATIVE TREATMENT

We take a medium without losses at the frequencies f_1 and f_2. We must write down Eq. (2.24) for $2f_1 = f_1 + f_1$ (set $f = 2f_1, f' = f_1$) and $f_1 = 2f_1 - f_1$ (set $f = f_1, f' = 2f_1$). This results in a coupled nonlinear system of differential equations for the complex amplitudes of the fundamental wave $\bar{E}(z,f_1)$ and the second harmonic $\bar{E}(z,2f_1)$, which are both plane waves:

$$\partial_z \bar{E}(z,2f_1) = -j \cdot \frac{\omega_1 \cdot d}{cn(2f_1)} \cdot \bar{E}^2(z,f_1) \cdot \exp(jz \cdot \Delta k)$$

$$(4.1)$$

$$\partial_z \bar{E}(z,f_1) = -j \cdot \frac{\omega_1 \cdot d}{cn(f_1)} \cdot \bar{E}(z,2f_1) \cdot \bar{E}^*(z,f_1) \cdot \exp(-jz \cdot \Delta k)$$

where we have set [see Eqs. (2.26) and (2.46)]

$$d := \chi_{\text{eff}}(-2f_1,f_1,f_1) = \left(\frac{1}{2}\right) \cdot \chi_{\text{eff}}(-f_1,2f_1,-f_1)$$

$$(4.2)$$

$$\Delta k := k(2f_1) - 2k(f_1) = \frac{2\omega_1}{c}[n(2f_1) - n(f_1)], \qquad k(-f_1) = -k(f_1)$$

d is real if there are no losses. Using Eq. A.6, one shows that the following relationship is fulfilled:

$$\frac{d}{dz}[n(2f_1)|\bar{E}(z,2f_1)|^2 + n(f_1)|\bar{E}(z,f_1)|^2] = 0$$

$$(4.3)$$

$$\langle S(z,f_1) \rangle + \langle S(z,2f_1) \rangle = \text{const}$$

The sum of the intensities at the frequencies f_1 and $2f_1$ is constant. The nonlinear medium does not take part in the energy exchange in the average; we have here a parametric process, (see Chapter 5). The exact solution of Eqs. (4.1) can be expressed using elliptic functions. However, we will treat now only two easily solvable special cases.

At first, we assume that the fundamental wave has such a high intensity that its (actual) decrease can be neglected: $\mathbf{E}(z,f_1) \approx \mathbf{E}(0,f_1) = \text{const}$. There is no pump depletion. This is valid as long as not too much power is transferred into the second harmonic (small signal approximation). Then the first ODE in Eq. (4.1) can be solved with the initial condition $\bar{E}(z = 0, 2f_1) = 0$:

$$\bar{E}(z,2f_1) = -j \cdot \frac{\omega_1 \cdot d \cdot z}{cn(2f_1)} \cdot \bar{E}^2(0,f_1) \cdot \frac{\exp(jz\Delta k) - 1}{j \cdot z\Delta k}$$

$$(4.4)$$

Thus, as intensity after a distance $z = L$,

$$\langle S(L, 2f_1) \rangle = \frac{2Z_0 \omega_1^2 \cdot d^2 \cdot L^2}{c^2 n^2(f_1) n(2f_1)} \cdot \langle S(0, f_1) \rangle^2 \cdot \frac{\sin^2(L\Delta k/2)}{(L\Delta k/2)^2} \qquad (4.5)$$

The intensity of the second harmonic is proportional to the square of the intensity of the fundamental wave, and as a function of the distance L, one has periodic behavior—as predicted in Section 4.1. Maxima occur for

$$L = L_k \left(m + \frac{1}{2} \right), \qquad m = 0, 1, 2, \ldots \qquad (4.6)$$

We denote here the distance between two neighboring maxima

$$L_k = \frac{2\pi}{\Delta k} = \frac{\pi c}{2\omega_1 [n(2f_1) - n(f_1)]} \qquad (4.7)$$

as *coherence length*. [At $\lambda = 1$ μm, one obtains with $n(2f_1) - n(f_1) \approx 0{,}01$ a coherence length of 50 μm.] With phase matching, one has $\Delta k = 0$, thus, according to Eq. (4.5), $\langle S(L, 2f_1) \rangle$ increases—for a constant pump wave— proportionally to L^2. A phase mismatch ($\Delta k \neq 0$) can then lead to a considerable reduction of intensity according to the function $(\sin u/u)^2$. In the literature, the coherence length sometimes is defined as $L_k' = L_k/2$.

In the second case, we assume phase matching [$\Delta k = 0, n(2f_1) = n(f_1)$] from the beginning. Then Eqs. (4.1) can easily be solved exactly, that is, with pump depletion included. As one can check immediately, the result is

$$\bar{E}(z, 2f_1) = -j \frac{\kappa c n(2f_1)}{\omega_1 d} \tanh \kappa z = -j\bar{E}(0, f_1) \tanh \kappa z$$

$$\bar{E}(z, f_1) = \frac{\kappa c n(2f_1)}{\omega_1 d} \cdot \frac{1}{\cosh \kappa z} = \bar{E}(0, f_1) \cdot \frac{1}{\cosh \kappa z} \qquad (4.8)$$

The initial conditions for this solution are

$$\bar{E}(0, 2f_1) = 0, \qquad \bar{E}(0, f_1) = \frac{\kappa c n(2f_1)}{\omega_1 d} \qquad (4.9)$$

This determines the value of the constant κ. We have chosen $\bar{E}(0, f_1)$ to be real; then $\bar{E}(z, f_1)$ also is real and $\bar{E}(z, 2f_1)$ imaginary. The power of the fundamental wave tends to 0 for $z \to \infty$, whereas the power of the second harmonic goes asymptotically to a maximum, if the whole power of the fundamental wave is converted.

However, in practice, there will be spontaneous scattering of photons with frequency $2f_1$ in a nonlinear medium; then fields at frequency f_1 are generated. Altogether, there will be again a periodic power exchange between the fundamental wave and the second harmonic, but this is not contained in the classical equations (4.1). Spontaneous processes can be described adequately only by quantum theory. But for induced processes, classical equations can be formulated whose predictions for strong fields agree with the expectation values calculated from the quantum theoretical equations of motions.

As SHG efficiency, we get from Eqs. (4.3) and (4.8), for $\bar{E}(0, 2f_1) = 0$,

$$\eta := \frac{\langle S(z, 2f_1) \rangle}{\langle S(0, f_1) \rangle} = \frac{\langle S(0, f_1) \rangle - \langle S(z, f_1) \rangle}{\langle S(0, f_1) \rangle}$$

$$= 1 - \frac{1}{\cosh^2 \kappa z} = \tanh^2 \left[\frac{\omega_1 d \bar{E}(0, f_1)}{cn(2f_1)} z \right]$$

Example 4.1   Let us consider nonlinear polarization for SHG in KDP. KDP is a negatively uniaxial crystal, with $3 = z$-axis as the optical axis. For phase-matched operation, both waves (fundamental and second harmonic) propagate along the same direction $\mathbf{k}$. The fundamental wave $\mathbf{E}$, is an ordinary wave, lying in the xy-plane; therefore, $\hat{E}_z(f_1) = 0$. Since for KDP, $\chi_{ijk} = 0$ if not all subscripts are different, the induced polarization has only the component [see Eq. (1.43)]

$$\hat{P}_z(2f_1) = \frac{\varepsilon_0}{2} \cdot \chi_{zxy} \cdot \hat{E}_x(f_1)\hat{E}_y(f_1)$$

If the projection of the $\mathbf{k}$-vector onto the xy-plane makes an angle φ with the x-axis, we have for the fundamental wave ($\mathbf{D}\|\mathbf{E}$ and $\mathbf{k} \perp \mathbf{E}$) $\hat{E}_x(f_1) = \hat{E}_1 \sin \varphi$, $\hat{E}_y(f_1) = -\hat{E}_1 \cos \varphi$. Thus,

$$\hat{P}_z(2f_1) = -\frac{\varepsilon_0}{4} \cdot \chi_{zxy} \cdot \sin 2\varphi \cdot \hat{E}_1^2$$

The produced second harmonic is an extraordinary wave, also propagating in the $\mathbf{k}$-direction. The electric field $\hat{E}_e$ associated with this e-wave lies in the $\mathbf{k}, z$-plane and is almost perpendicular to $\mathbf{k}$ ($\mathbf{D}$ would be perpendicular to $\mathbf{k}$, but $\mathbf{D}$ and $\mathbf{E}$ are almost parallel, since the ellipsoid does not deviate much from a sphere). With θ as the angle between $\mathbf{k}$ and the optical axis, the component of $\hat{P}_z(2f_1)$ in the direction of the excited second harmonic $\hat{E}_e$ is given by

$$\hat{P}_{NL}(2f_1) = \hat{P}_z(2f_1) \cos \left(\frac{\pi}{2} - \theta \right) = -\frac{\varepsilon_0}{4} \cdot \chi_{zxy} \cdot \sin \theta \sin 2\varphi \cdot \hat{E}_1^2 \qquad \blacksquare$$

4.3   SUPPLEMENTS

4.3.1   Gaussian Beams

Up to now, we have used plane waves. More realistic is the case that the fundamental is a Gaussian TEM_{00} mode of the form

$$\widehat{\phi}_{00} = \sqrt{\frac{4k_0 P_{00}}{\pi b}} \cdot \frac{q(0)}{q(z)} \cdot \exp\left[-jk_0 n \cdot \frac{x^2 + y^2}{2q(z)}\right] \cdot \exp(-jk_0 n \cdot z)$$

$$= \sqrt{\frac{4k_0 P_{00}}{\pi b}} \cdot \frac{w_0}{w(z)} \exp\left[-\frac{x^2 + y^2}{w^2(z)} - jk_0 n \frac{x^2 + y^2}{2R(z)} - jk_0 n \cdot z + j\psi(z)\right] \quad (4.10)$$

$$= \overline{\phi}_{00} \exp(-jk_0 n \cdot z)$$

$$q(z) = z + \frac{jb}{2}, \quad w^2(z) = w_0^2\left(1 + \frac{4 \cdot z^2}{b^2}\right), \quad b = k_0 n w_0^2, \quad \frac{1}{q} = \frac{1}{R} - \frac{2j}{k_0 n w^2}$$

with w_0 as the spot radius in the beam waist, b as the confocal parameter, $R(z)$ as the beam radius, $k_0 = \omega_1/c$, $n = n(f_1)$, $\psi = \arctan 2z/b$, and the beam waist lies at $z = 0$. According to Eq. (A.7), we use as power P_{00} the intensity integrated over the total cross section:

$$P_{00} = \iint \langle S \rangle dx\, dy = \iint \left(\frac{1}{2}\right) \cdot n(f_1) \cdot |\widehat{\phi}_{00}(x, y, z = 0, f_1)|^2 dx\, dy \quad (4.11)$$

For crystal lengths $L \ll b$, one can use rather well plane waves (the beam cross section remains approximately constant), and we get due to the concentration of intensity toward the beam axis an increase of the SHG conversion efficiency. If, however, $L \gg b$, the increase in beam cross section reduces the efficiency. Thus, an optimal value for $L(L \sim b)$ should exist.

We set

$$P(L, 2f_1) := \iint \langle S(x, y, L, 2f_1) \rangle dx\, dy \quad (4.12)$$

Integrate $\langle S(x, y, z = 0, f_1) \rangle^2$ over the cross section using Eqs. (4.11) and (4.10) and obtain

$$\frac{P_{00}^2}{\pi w_0^2} = \iint \langle S(x, y, z = 0, f_1) \rangle^2 dx\, dy \quad (4.13)$$

For the case in which the fundamental wave is incident as a Gaussian TEM_{00} mode, we find using Eq. (4.4) (which is indeed not quite correct since the formula there was derived for plane waves)

$$P(L, 2f_1) = \frac{2Z_0\omega_1^2 \cdot d^2}{c^2 n^2(f_1)n(2f_1)} \cdot \frac{\mathcal{P}_{00}^2}{\pi w_0^2} L^2 \cdot \frac{\sin^2(L\Delta k/2)}{(L\Delta k/2)^2} \tag{4.14}$$

Again, we obtain periodic behavior and for $\Delta k = 0$ an increase proportional to L^2. The exact theory yields instead

$$P(L, 2f_1) = \frac{2Z_0\omega_1^2 d^2}{c^2 n^2(f_1)n(2f_1)} \cdot \frac{\mathcal{P}_{00}^2}{\pi w_0^2} \cdot L \cdot b \cdot e^{-\alpha L} \cdot h \tag{4.15}$$

α is a measure for the absorption in the crystal; h is a factor reducing the SHG conversion efficiency with a maximum value of 1.068 [at $L = 2.84 \times b$, $90°$ phase matching, lossless case with $\alpha = 0$, beam waist in the middle of the crystal, wave vector mismatch $\Delta k = 3.2/L$], that is,

$$\mathcal{P}_{max}(L, 2f_1) = \frac{2Z_0\omega_1^2 d^2}{c^2 n^2(f_1)n(2f_1)} \cdot \frac{\mathcal{P}_{00}^2}{\pi w_0^2} \cdot b^2 \cdot 2.84 \cdot 1.068 \tag{4.16}$$

One would get the same maximum value from Eq. (4.14), if one takes there $\Delta k = 0$ and chooses $L = 1.74 \cdot b$. Thus, we find $L \approx 2b$.

4.3.2 SHG in a Resonator

The power $\mathcal{P}(L, 2f_1)$ increases quadratically with crystal length L only if L is in the vicinity of $2b$. Therefore, it is reasonable to use a resonator for the second harmonic. The fundamental wave will pass through the resonator from left to right (Fig. 4.1). The mirrors are transparent for the fundamental wave; for the

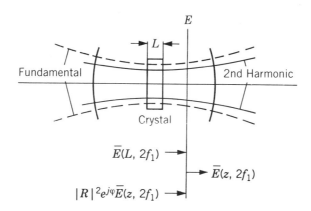

FIGURE 4.1 SHG in a resonator. (Reproduced with permission from G. K. Grau, *Quantenelektronik*, Vieweg-Verlag Braunschweig, 1978.)

second harmonic, they show a reflection factor $R(|R| \approx 1)$ and a transmission factor T (for lossless mirrors, we have $|R|^2+|T|^2 = 1$). With each passage from left to right, the field of the generated second harmonic induces a reduction of the power of the fundamental wave. This is connected with a generation of a contribution $\bar{E}(L, 2f_1)$ at $2f_1$ according to Eq. (4.3). φ is the phase shift of the field of the second harmonic in the resonator for a whole round-trip from the plane E back again to the plane E (the phase shifts due to reflections at the mirrors are already contained in it). In the stationary case, we have the following balance between the incoming and outgoing waves in the plane E (at position z):

$$\bar{E}(z, 2f_1) = \bar{E}(L, 2f_1) + |R|^2 e^{j\varphi} \bar{E}(z, 2f_1)$$

or

$$\bar{E}(z, 2f_1) = \frac{\bar{E}(L, 2f_1)}{1 - |R|^2 \exp(j\varphi)} \tag{4.17}$$

Through the right mirror, the amplitude

$$\bar{E}_t(2f_1) = \bar{E}(z, 2f_1) \cdot T \cdot e^{j\Phi} \tag{4.18}$$

is transmitted, where Φ takes into account the phase shift from the plane E to the mirror. Substituting Eq. (4.17) in Eq. (4.18), calculating the intensity according to Eq. (A.6), and integrating over the cross section, we find for the power of the second harmonic transmitted through the mirror

$$\mathcal{P}_t(2f_1) = \frac{|T|^2 \mathcal{P}(L, 2f_1)}{1 + |R|^4 - 2|R|^2 \cos \varphi} \tag{4.19}$$

and with a correct adjustment of the resonator length ($\varphi = 2\pi m$),

$$\mathcal{P}_t(2f_1) = \frac{|T|^2 \mathcal{P}(L, 2f_1)}{(1 - |R|^2)^2} = \frac{\mathcal{P}(L, 2f_1)}{|T|^2} \tag{4.20}$$

For $\mathcal{P}(L, 2f_1)$, one can use in the most favorable case the power derived from Eq. (4.16). For $|R|^2 = 0.99$ and $|T|^2 = 0.01$, this gives rise to a factor 100, compared with the nonresonant case.

4.3.3   Laser with Internal Frequency Conversion

See Fig. 4.2. A laser oscillates at a frequency f. The mirror S_1 is highly reflecting for signals at the frequencies f and $2f$; the mirror S_2 is highly reflecting only for signals at the frequency f, but nearly transparent for signals at the frequency $2f$. Some portion of the power $\mathcal{P}_{00}$ circulating in the resonator at the frequency f in both directions in the nonlinear crystal will be transferred into power $\mathcal{P}(L_c, 2f)$ at the frequency $2f$ after passing through the crystal of length L_c. In reality, one

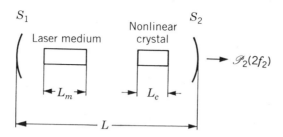

FIGURE 4.2   Laser with internal frequency conversion.

should add the fields; however, without stabilizing the resonator components, one averages effectively over all phases and obtains thus from Eq. (4.14) for the outgoing power $\mathcal{P}_2$ at the frequency $2f$ a proportionality with $\mathcal{P}_{00}^2$:

$$\mathcal{P}_2 = 2c_\kappa \mathcal{P}_{00}^2 \tag{4.21}$$

As a consequence of the conversion and the coupling of the converted power through the mirror, there are losses for the fundamental wave that can be attributed to a loss parameter δ_κ characterizing the losses per passage:

$$\delta_\kappa = \frac{\mathcal{P}_2}{\mathcal{P}_{00}} = 2c_\kappa \mathcal{P}_{00} \tag{4.22}$$

The remaining losses per passage are characterized by δ. In the stationary case, the large signal amplification

$$\alpha_{LS} = \frac{2L_m \alpha(f, f_{32})}{1 + 2\mathcal{P}_{00}/[AS_s(f)]}$$

[α = power amplification constant with midfrequency f_{32}, A = cross-sectional area of the laser beam in the amplifying medium of length L_m, $S_s(f)$ = saturation parameter, the factor of 2 in the denominator stems from taking into account the forward and backward running wave] for a laser with a homogeneous spectral line must equal the total losses:

$$\alpha_{LS} = \delta + \delta_\kappa = \delta + 2c_\kappa \mathcal{P}_{00}$$

Here, we eliminate now $\mathcal{P}_{00}$ using Eq. (4.21) and solve for $\mathcal{P}_2$. We find the optimal coupling if we set $\partial \mathcal{P}_2/\partial c_\kappa = 0$ and with

$$c_{\kappa,\text{opt}} = \frac{\delta}{[AS_s(f)]} \tag{4.23}$$

obtain

$$P_{2,\text{max}} = \left(\frac{1}{2}\right) \cdot AS_s(f) \left[\sqrt{2L_m\alpha(f,f_{32})} - \sqrt{\delta}\right]^2$$

(4.24)

$$P_{00,\text{opt}} = \sqrt{\frac{P_{2,\text{max}}}{2c_{\kappa,\text{opt}}}} = \frac{AS_s(f)}{2\sqrt{\delta}} \cdot \left[\sqrt{2L_m\alpha(f,f_{32})} - \sqrt{\delta}\right]$$

By using this internal SHG in a Nd:YAG laser together with a nonlinear $Ba_2Na(NbO_3)_5$ crystal, TEM_{00} beams at $\lambda = 0.53$ μm, the powers of which are in the order of magnitude of 1 W can be achieved (in the continuous case).

PROBLEMS

Section 4.2

4.1 Derive Eq. (4.3).

4.2 Derive Eq. (4.8).

4.3 Using a field $\bar{E}(0,f_1) = 7 \times 10^6$ V/m, $n \approx 1.5$ at a wavelength of $\lambda = 1.06$ μm (Nd:YAG laser), $d \approx 2.5 \times 10^{-12}$ m/V for KDP, and a crystal of length $L = 1$ cm, calculate the SHG conversion efficiency.

4.4 Compare the SHG efficiencies in the case of no pump depletion for perfect phase matching ($\Delta k = 0$) after $L = 1$ cm and for nonphase-matched operation after an optimal length $L = L_k(m + 1/2)$. How large is the reduction, if the phase-matching condition is not fulfilled? Take the numerical values of the example after Eq. (4.7).

Section 4.3

4.5 Show that the Gaussian beam $\bar{\phi}_{00}$, Eq. (4.10), is a solution of the paraxial Helmholtz equation (see Problem 2.3).

4.6 A 1-mW He-Ne laser produces a Gaussian beam of wavelength $\lambda = 633$ nm and a spot-size radius $w_0 = 0.05$ mm ($n = 1$). (Reprinted with permission from Saleh and Teich, 1991.)

    (a) Determine the angular divergence of the beam [derived from the beam width $w(z)$ for large values of z] and its diameter at $z = 3.5 \times 10^5$ km (distance to the moon).

    (b) What is the radius of curvature of the wavefront at $z = 0, z = b/2, z = b$, and $z = 2b$?

    (c) What is the optical intensity on the axis ($x = y = 0$) at the beam center ($z = 0$) and at $z = b$? Compare this result with the intensity at $z = b$ of a

100-W spherical wave produced by a small isotropically emitting light source at $z = 0$.

BIBLIOGRAPHY

R. W. Boyd. *Nonlinear Optics.* Academic Press, San Diego, CA, 1992.

P. N. Butcher. *Nonlinear Optical Phenomena.* Bulletin 200, Engineering Experimental Station, Ohio State University, Columbus, Ohio.

G. K. Grau. *Quantenelektronik.* Vieweg, Braunschweig, 1978.

D. L. Mills. *Nonlinear Optics.* Springer-Verlag, Berlin, 1991.

M. Schubert and B. Wilhelmi. *Nonlinear Optics and Quantum Electronics.* Wiley, New York, 1986.

B. E. A. Saleh and M. C. Teich. *Fundamentals of Photonics.* Wiley, New York, 1991.

Y. R. Shen. *The Principles of Nonlinear Optics.* Wiley, New York, 1984.

A. Yariv. *Quantum Electronics.* Wiley, New York, 1975.

CHAPTER FIVE

Parametric Effects

In *parametric processes*, energy and momentum conservation must be satisfied alone by the photons. The nonlinear medium is, at least at the end of the process, in the same state as in the beginning. The ideal parametric process is lossless and χ is real. (In fact, it is not possible to completely avoid losses caused by the medium.) In general, parametric processes are processes in which a parameter of an energy reservoir of an oscillator system is varied periodically. As a consequence, different oscillating modes of the light field are coupled together periodically or non periodically. Of course, parametric processes occur in not only three-wave mixing but also four-wave mixing.

5.1 THE BASIC EQUATIONS

From Eq. (2.24), we obtain for the frequencies f_1, f_2, f_3 with $f_3 = f_1 + f_2$ in turn

$$\partial_z \bar{E}(z,f_3) = -\frac{j\omega_3\chi}{2cn(f_3)} \cdot \bar{E}(z,f_1) \cdot \bar{E}(z,f_2) \cdot \exp(+j \cdot \Delta k \cdot z) \qquad (5.1)$$

$$\partial_z \bar{E}(z,f_2) = -\frac{j\omega_2\chi^*}{2cn(f_2)} \cdot \bar{E}(z,f_3) \cdot \bar{E}^*(z,f_1) \cdot \exp(-j \cdot \Delta k \cdot z) \qquad (5.2)$$

$$\partial_z \bar{E}(z,f_1) = -\frac{j\omega_1\chi^*}{2cn(f_1)} \cdot \bar{E}(z,f_3) \cdot \bar{E}^*(z,f_2) \cdot \exp(-j \cdot \Delta k \cdot z) \qquad (5.3)$$

Here, we have set, according to Eq. (2.25), in the lossless case

$$\chi := \chi_{\text{eff}}(-f_3,f_1,f_2) = \chi_{\text{eff}}^*(-f_2,f_3,-f_1) = \chi_{\text{eff}}^*(-f_1,f_3,-f_2) \qquad (5.4)$$

and

$$\Delta k = k(f_3) - k(f_1) - k(f_2) = \frac{2\pi}{c} \cdot [f_3 n(f_3) - f_1 n(f_1) - f_2 n(f_2)] \qquad (5.5)$$

$k(-f) = -k(f)$. Quite generally, the intensities, Eq. (A.6), satisfy here

$$\frac{\partial}{\partial z}[\langle S(z,f_1)\rangle + \langle S(z,f_2)\rangle + \langle S(z,f_3)\rangle] = 0 \qquad (5.6)$$

as one can show using Eqs. (5.1) through (5.3). The sum of the intensities of all three light waves is, in fact, constant; the medium does not take part in the energy exchange in the mean.

As already mentioned, SHG is also a parametric process, for which we can set $f_1 = f_2$ and replace

$$\langle S(z,f_1)\rangle + \langle S(z,f_2)\rangle = 2\langle S(z,f_1)\rangle \Rightarrow \langle S(z,f_1)\rangle$$

See Eq. (4.3). The general solution of the system of differential equations (5.1) through (5.3) can again be expressed with elliptic functions. However, we shall discuss here only the case of phase matching $\Delta k = 0$. The condition $\Delta k = 0$ can be fulfilled in a crystal. Figure 5.1 shows some parametric processes.

5.2   SUM AND DIFFERENCE FREQUENCY MIXING

The intensities of the waves at the frequencies f_1, f_2 are assumed to be sufficiently high enough so that they may be approximated by the very large constant

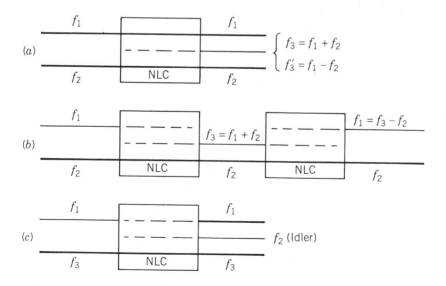

FIGURE 5.1   Parametric processes in a nonlinear crystal NLC (*a*) sum and difference frequency, (*b*) up and down conversion (*c*) parametric amplifier.

amplitudes $\bar{E}(0, f_1)$, $\bar{E}(0, f_2)$. With $\Delta k = 0$, Eq. (5.1) yields

$$\bar{E}(L, f_3) = -\frac{j\omega_3 \chi}{2cn(f_3)} \cdot \bar{E}(0, f_1)\bar{E}(0, f_2) \cdot L \qquad (5.7)$$

that is, the amplitude of the wave at the sum frequency f_3 grows proportionally to the interaction length L if phase-matched. In sum frequency mixing, one photon at the frequency f_1 and one photon at f_2 are annihilated by two-photon absorption and the atom jumps into a higher (virtual) energy level. Subsequently, a photon $f_3 = f_1 + f_2$ is generated (emitted), and the atom jumps back into the original state. With sum frequency mixing of two lasers in the visible spectral domain (one at a fixed frequency, the other is tunable), a tunable light source in the UV can be produced. Equation (5.7) has exactly the same structure as Eq. (4.4) in the case of SHG for $\Delta k = 0$ ($f_1 = f_2 = f_3/2$).

If, however, the waves at f_2 and f_3 are very large and approximately constant, one can use the amplitudes $\bar{E}(0, f_2)$, $\bar{E}(0, f_3)$ at $z = 0$. From Eq. (5.3), it then follows that

$$\bar{E}(L, f_1) = -\frac{j\omega_1 \chi^*}{2cn(f_1)} \cdot \bar{E}(0, f_3)\bar{E}^*(0, f_2) \cdot L \qquad (5.8)$$

A wave at the difference frequency $f_1 = f_3 - f_2$ evolves. Here, at first, a photon with frequency ω_3 is absorbed. Due to a two-photon emission, photons at the energies $\hbar\omega_1$ and $\hbar\omega_2$ are generated. ω_2 photons that are initially already present stimulate the process. Difference frequency mixing is used for the tunable generation of coherent radiation in the far infrared. For example, with a CO_2 laser, producing coherent radiation at about 100 lines with wavelengths between 9.1 and 11.0 μm, one can generate about 5000 lines with wavelengths between 70 μm and 2 mm with a mean distance of 0.04 cm^{-1} by the difference frequency mixing of these frequencies in GaAs.

If initially (at $z = 0$), there are photons of two fields at f_2 and f_1 ($> f_2$), a difference frequency $f_3' = f_1 - f_2$ or a sum frequency $f_3 = f_1 + f_2$ in principle can develop (see Fig. 5.1a). The decision will be made by momentum balance (phase matching), which can be adjusted by an appropriate orientation of the crystal: If $k_1 - k_2 = k_3'$ holds together with $f_3' = f_1 - f_2$, one has difference frequency generation. On the other hand, $k_1 + k_2 = k_3$ with $f_3 = f_1 + f_2$ yields sum frequency generation. Actually, waves with the frequencies f_3 and f_3' are simultaneously generated locally. But only that wave can develop which experiences a constructive interference due to the correct phase matching.

5.3   FREQUENCY UP AND DOWN CONVERSION

If only $\bar{E}(z, f_2)$ is very large and approximately equal to $\bar{E}(0, f_2)$ (that means a constant pump power $\mathcal{P}_2$), then it follows from Eqs. (5.1) and (5.3) that

$$\frac{\partial}{\partial z}\bar{E}(z,f_3) = -\frac{j\omega_3\chi}{2cn(f_3)} \cdot \bar{E}(0,f_2)\bar{E}(z,f_1)$$

$$\frac{\partial}{\partial z}\bar{E}(z,f_1) = -\frac{j\omega_1\chi^*}{2cn(f_1)} \cdot \bar{E}^*(0,f_2)\bar{E}(z,f_3)$$

(5.9)

or, after elimination of $\bar{E}(z,f_1)$ and $\bar{E}(z,f_3)$, respectively;

$$\left(\frac{\partial^2}{\partial z^2} + \kappa^2\right)\begin{bmatrix} \bar{E}(z,f_3) \\ \bar{E}(z,f_1) \end{bmatrix} = 0$$

with solutions of the form $\cos(\kappa z)$ and $\sin(\kappa z)$. Here,

$$\kappa^2 = \frac{\omega_1\omega_3|\chi|^2|\bar{E}(0,f_2)|^2}{4c^2n(f_1)n(f_3)}$$

(5.10)

Explicitly, the solution becomes

$$\begin{bmatrix} \bar{E}(z,f_3) \\ \bar{E}(z,f_1) \end{bmatrix} = \begin{bmatrix} \cos\kappa z & -\dfrac{j\omega_3\chi\bar{E}(0,f_2)}{2cn(f_3)\kappa} \cdot \sin\kappa z \\ -j\dfrac{2cn(f_3)\kappa}{\omega_3\chi\bar{E}(0,f_2)}\sin\kappa z & \cos\kappa z \end{bmatrix} \cdot \begin{bmatrix} \bar{E}(0,f_3) \\ \bar{E}(0,f_1) \end{bmatrix}$$

(5.11)

For $\bar{E}(0,f_3) = 0$, $\bar{E}(0,f_1) \neq 0$, we get frequency up conversion

$$\bar{E}(z,f_3) = -\frac{j\omega_3\chi\bar{E}(0,f_2)}{2cn(f_3)\kappa} \cdot \bar{E}(0,f_1) \cdot \sin\kappa z$$

$$\bar{E}(z,f_1) = \bar{E}(0,f_1) \cdot \cos\kappa z$$

(5.12)

$|\bar{E}(z,f_3)|$ is increasing, controlled by the amplitude $\bar{E}(0,f_1)$, and reaches a maximum value after a distance $L = \pi/(2\kappa)$; here, we have $\bar{E}(L,f_1) = 0$. Now, the process of frequency down conversion begins. One can get it directly from the initial conditions $\bar{E}(0,f_1) = 0, \bar{E}(0,f_3) \neq 0$. There is then a periodic power exchange. Figure 5.1b shows at the left the up converter and at the right the down converter, with the length of the nonlinear crystal $L = \pi/(2\kappa)$ in both cases.

As a matter of fact, the wave at the frequency f_2 must also take part in this power exchange because of Eq. (5.6). We obtain from Eq. (5.12) for the intensity in up conversion with $\bar{E}(0, f_3) = 0$

$$\frac{\langle S(z, f_1)\rangle}{\omega_1} + \frac{\langle S(z, f_3)\rangle}{\omega_3} = \frac{\langle S(0, f_1)\rangle}{\omega_1} + \left[\frac{\langle S(0, f_3)\rangle}{\omega_3}\right] = \text{const} = C_1 \qquad (5.13)$$

and in down conversion with $\bar{E}(0, f_1) = 0$

$$\frac{\langle S(z, f_1)\rangle}{\omega_1} + \frac{\langle S(z, f_3)\rangle}{\omega_3} = \frac{\langle S(0, f_3)\rangle}{\omega_3} + \left[\frac{\langle S(0, f_1)\rangle}{\omega_1}\right] = \text{const} = C_2 \qquad (5.14)$$

From Eqs. (5.13) and (5.14), it follows for both up and down conversion that

$$\frac{1}{\omega_1} \cdot \frac{\partial \langle S(z, f_1)\rangle}{\partial z} = -\frac{1}{\omega_3} \cdot \frac{\partial \langle S(z, f_3)\rangle}{\partial z} \qquad (M_1 = -M_3) \qquad (5.15)$$

Eliminating $\langle S(z, f_1)\rangle$ with Eq. (5.6), we get

$$\frac{1}{\omega_2} \cdot \frac{\partial \langle S(z, f_2)\rangle}{\partial z} = -\frac{1}{\omega_3} \cdot \frac{\partial \langle S(z, f_3)\rangle}{\partial z} \qquad (M_2 = -M_3) \qquad (5.16)$$

and from the two preceding equations finally

$$\frac{1}{\omega_1} \cdot \frac{\partial \langle S(z, f_1)\rangle}{\partial z} = \frac{1}{\omega_2} \cdot \frac{\partial \langle S(z, f_2)\rangle}{\partial z} \qquad (M_1 = +M_2) \qquad (5.17)$$

These are again the Manley–Rowe equations for parametric processes; see Eq. (2.8). In up conversion, the wave at f_2 contributes to each energy portion $\hbar\omega_1$ an energy $\hbar\omega_2$; the result is the generation of a quantum at the frequency $f_3 = f_1 + f_2$. According to Eqs. (5.15) and (5.16), $\langle S(z, f_3)\rangle$ increases at the expense of $\langle S(z, f_1)\rangle$ and $\langle S(z, f_2)\rangle$. In down conversion, an energy quantum $\hbar\omega_3$ is split into $\hbar\omega_1$ and $\hbar\omega_2$ with the help of an energy quantum $\hbar\omega_2$ (this is an induced or stimulated process).

As conversion efficiency in up conversion, one finds

$$\eta := \frac{\langle S(L, f_3)\rangle}{\langle S(0, f_1)\rangle} = \frac{\omega_3}{\omega_1} \sin^2 \kappa L \approx \frac{\omega_3^2 |\chi|^2 |\bar{E}(0, f_2)|^2}{4c^2 n(f_1) n(f_3)} L^2 \sim \frac{\mathcal{P}_2}{A} \cdot L^2$$

The approximation holds for $\kappa L \ll 1$. Thus, η depends on the pump power $\mathcal{P}_2$ and the ratio L^2/A, of the square of the length of the nonlinear medium to the cross section A. For a numerical value, see Problem 5.4.

Up conversion is, in a sense, a special case of sum frequency generation: Two frequencies f_1 and f_2 are mixed additively. We only assume here $|\bar{E}(z, f_2)| \gg |\bar{E}(z, f_1)|$, whereas the fields in sum frequency mixing are of the same order of magnitude usually. Correspondingly, the down conversion is difference mixing with $|\bar{E}(z, f_2)| \gg |\bar{E}(z, f_3)|$. Since in contrast to sum frequency, we have now allowed for pump depletion at f_1 (not f_2), we obtain here the periodic increasing and decreasing of the fields at f_1 and f_3.

An important application of parametric up conversion lies in the detection of far infrared radiation (where detectors have a lower efficiency or are very slow or need considerable cooling): A signal at a lower frequency f_1 is mixed with a strong laser beam at f_2. This yields a higher frequency $f_3 = f_1 + f_2$ in the near infrared or visible spectrum, which is more easily detectable.

5.4   PARAMETRIC AMPLIFICATION

Up to now, we did not speak of amplification; amplification means control of an increasing wave at a frequency f' by a small initial amplitude at the same frequency f'. Quite generally, an amplifier is a device through which a large amount of power is controlled by a small amount of power.

In Eqs. (5.1) through (5.3) $\bar{E}(z, f_3)$ is now the pump wave, very large and approximately equal to $\bar{E}(0, f_3)$. Then, Eqs. (5.2) and (5.3) yield after elimination of $\bar{E}(z, f_2)$ and $\bar{E}(z, f_1)$, respectively,

$$\left(\frac{\partial^2}{\partial z^2} - \kappa^2\right)\left[\begin{array}{c} \bar{E}(z, f_1) \\ \bar{E}(z, f_2) \end{array}\right] = 0 \qquad (5.18)$$

with solutions of the form $\cosh(\kappa z)$ and $\sinh(\kappa z)$. Here,

$$\kappa^2 = \frac{\omega_1 \omega_2 |\chi|^2 |\bar{E}(0, f_3)|^2}{4c^2 n(f_1) n(f_2)} \qquad (5.19)$$

Explicitly, the solution is

$$\left[\begin{array}{c} \bar{E}(z, f_1) \\ \bar{E}^*(z, f_2) \end{array}\right] = \left[\begin{array}{cc} \cosh \kappa z & -\dfrac{j\omega_1 \chi^* \bar{E}(0, f_3)}{2cn(f_1)\kappa} \sinh \kappa z \\ j\dfrac{2cn(f_1)\kappa}{\omega_1 \chi^* \bar{E}(0, f_3)} \sinh \kappa z & \cosh \kappa z \end{array}\right]\left[\begin{array}{c} \bar{E}(0, f_1) \\ \bar{E}^*(0, f_2) \end{array}\right]$$

$$(5.20)$$

Assume $\bar{E}(0, f_2) = 0$; $\bar{E}(0, f_1) \neq 0$ is a small input signal that increases thus proportionally to $\bar{E}(0, f_1)$ as $\cosh \kappa z$. At the same time, a wave (the idler wave) is produced at the frequency f_2 that increases proportionally to $\sinh(\kappa z)$. The power for both waves comes, of course, from the pump wave with the highest frequency $\omega_3 = \omega_1 + \omega_2$. A few initially existing energy quanta $\hbar\omega_1$ of the signal are sufficient to stimulate the decay of the pump quantum $\hbar\omega_3$ into two portions $\hbar\omega_1$ and $\hbar\omega_2$. $\hbar\omega_1$ contributes to coherent amplification of the signal; $\hbar\omega_2$ constitutes the idler wave (see Fig. 5.1c). Similarly, a photon at ω_2 induces again the emission of a photon at ω_1, and so on. This buildup process in parametric amplification is responsible for an exponential increase of both waves at ω_1 and ω_2 [as long as $\bar{E}(z, f_3)$ can be considered constant]. The amplification of the signal wave is given by

$$\frac{\langle S(L, f_1) \rangle}{\langle S(0, f_1) \rangle} = \frac{|\bar{E}(L, f_1)|^2}{|\bar{E}(0, f_1)|^2} = \cosh^2 \kappa L \approx \left(\frac{1}{4}\right) \exp(2\kappa L) \qquad \text{for } \kappa L \gg 1 \quad (5.21)$$

with the (amplitude) amplification constant (per unit length) κ. Even for large pump power densities, κ remains small. This is the reason why the effect is used mainly to generate oscillations in a parametric oscillator. In this case, we still need feedback (supplied by a resonator), and the gain must cancel the losses.

5.5 PARAMETRIC OSCILLATOR

For the doubly resonant parametric oscillator (DRO), a nonlinear crystal (length L) is equipped with two mirrors S_1, S_2 at the end faces, which are perfectly transparent at the pump frequency f_3 (see Fig. 5.2). The mirror S_1 is perfectly reflecting at f_1 (signal) and f_2 (idler) with the following amplitude reflection coefficients:

$$R'_1 = R'(f_1) = 1, \quad R'_2 = R'(f_2) = 1, \qquad R'_3 = R'(f_3) = 0$$

but the mirror S_2 is only partially transparent. The reflection coefficients at S_2 are very large, but smaller than 1:

$$R_1 = R(f_1) = \exp\left(\frac{-\delta_1}{2}\right), \quad R_2 = R(f_2) = \exp\left(\frac{-\delta_2}{2}\right), \quad R_3 = R(f_3) = 0$$

The losses of the resonator at f_1 and f_2 stem from the radiation losses through S_2; they are described by the power loss parameters per round-trip δ_1, δ_2. The quality

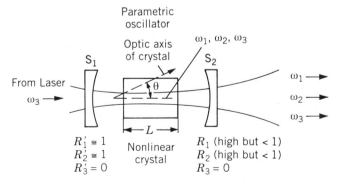

FIGURE 5.2 Parametric oscillator. (Reproduced with permission from A. Yariv, *Quantum Electronics*, John Wiley & Sons, Inc., New York, 1975.)

(Q) factor is assumed to be large enough so that inside the resonator, the amplitudes $\bar{E}(f_1)$, $\bar{E}(f_2)$ can be considered independent of z. Since both backward running waves f_1 and f_2 in the resonator produce a signal at frequency f_3 via sum frequency mixing, an apparent pump reflection develops with an amplitude $\bar{E}_r(f_3)$. The losses through mirror S_2 can be described by a distributed loss mechanism on the way from left to right:

$$\bar{E}(L, f_1) = \bar{E}(0, f_1) \cdot \exp\left(\frac{-\delta_1}{2}\right), \qquad \bar{E}(L, f_2) = \bar{E}(0, f_2) \cdot \exp\left(\frac{-\delta_2}{2}\right)$$

After inserting in Eq. (5.20), with $z = L$, we obtain

$$\begin{bmatrix} \cosh \kappa L - \exp\left(\dfrac{-\delta_1}{2}\right) & -\dfrac{j\omega_1 \chi^* \bar{E}(0, f_3)}{2cn(f_1)\kappa} \sinh \kappa L \\[2mm] j\dfrac{2cn(f_1)\kappa}{\omega_1 \chi^* \bar{E}(0, f_3)} \sinh \kappa L & \cosh \kappa L - \exp\left(\dfrac{-\delta_2}{2}\right) \end{bmatrix} \cdot \begin{bmatrix} \bar{E}(0, f_1) \\[2mm] \bar{E}^*(0, f_2) \end{bmatrix} = 0 \quad (5.22)$$

A nontrivial solution exists only if the determinant of coefficients vanishes:

$$\exp\left(\frac{-\delta_1}{2}\right) \exp\left(\frac{-\delta_2}{2}\right) - \cosh \kappa L \left[\exp\left(\frac{-\delta_1}{2}\right) + \exp\left(\frac{-\delta_2}{2}\right) \right] + 1 = 0$$

With the approximations $\delta_1 \ll 1, \delta_2 \ll 1$, $R_i \approx 1 - \delta_i/2$, and $\cosh \kappa L \approx 1 + \kappa^2 L^2/2$ (small amplification κ), while the term

$$\frac{\delta_1 + \delta_2}{2} \cdot \frac{\kappa^2 L^2}{2}$$

is neglected, the following threshold condition results:

$$4\kappa^2 L^2 = \delta_1 \delta_2, \qquad 2\kappa L = \sqrt{\delta_1 \delta_2} \tag{5.23}$$

The parametric gain equals the geometric mean of the losses from the signal and the idler. As threshold intensity, we find using $|\bar{E}(0, f_3)|^2$ from Eq. (5.19) and with $n_i = n(f_i)$

$$\langle S(f_3) \rangle_{\text{th}} = \frac{n_3}{2Z_0} |\bar{E}(0, f_3)|^2 = \frac{4c^2}{2Z_0} \cdot \frac{n_1 n_2 n_3}{\omega_1 \omega_2 |\chi|^2} \cdot \kappa^2 = \frac{c^2}{2Z_0} \cdot \frac{n_1 n_2 n_3 \delta_1 \delta_2}{\omega_1 \omega_2 |\chi|^2 L^2}. \tag{5.24}$$

Example 5.1 If LiNbO$_3$ is used as nonlinear crystal, we find for $\lambda_1 = \lambda_2 = 1$ μm, $L = 1$ cm, $\delta_1 \approx \delta_2 \approx 4 \times 10^{-2}$ (4% loss per passage in the power), $n(f_1) = n(f_2) = n(f_3) = 1.5$, and $|\chi|^2 = 5 \times 10^{-23}$ (m/V)^2, a value of $\langle S(f_3) \rangle_{\text{th}} = 3.63 \times 10^3$ W/cm^2. This can easily be achieved (even continuously). ∎

Thus, the optical parametric oscillator is of importance in practice for converting a laser pump power at f_3 into coherent radiation at the signal and idler frequency.

Decreasing δ_1, δ_2 (increasing R_1, R_2) and operating the oscillator nearer to the degenerate case (the degenerate case is characterized by $f_1 = f_2 = f_3/2$) results in the decrease in the threshold power since the factor $f_1 f_2 = f_1(f_3 - f_1)$ is maximum for $f_1 = f_3/2$.

For the singly resonant oscillator (SRO), for which, for example, only the signal f_1 is resonant, the threshold power is higher since δ_2 is very large $[R(f_2) \approx 0]$.

Example 5.2   An alternative derivation of the threshold pump intensity for the parametric oscillator is as follows.

If we rewrite Eqs. (5.2) and (5.3) with the new variables

$$a_i(z) = \sqrt{\frac{n_i}{\omega_i}} \cdot \bar{E}(z, f_i)$$

and assume perfect phase matching ($\Delta k = 0$) and an undepleted pump, $\bar{E}(z, f_3) = \bar{E}(0, f_3)$, we find

$$a_i'(z) = -\left(\frac{j}{2}\right)\gamma a_k^*(z), \qquad i, k = 1, 2, \qquad i \neq k$$

where

$$\gamma = \frac{\chi}{c}\sqrt{\frac{\omega_1\omega_2\omega_3}{n_1 n_2 n_3}} \cdot a_3(0) = 2\kappa$$

Now, losses have to be included, which are caused by scattering and absorption in the nonlinear medium and by losses through the mirrors of the resonator. We write therefore

$$a_i'(z) = -\left(\frac{\alpha_i}{2}\right)a_i - \left(\frac{j}{2}\right)\gamma a_k^*$$

These power attenuation coefficients α_i have the dimension of an inverse length. In the steady state, all derivatives are zero, and we write the system as

$$\frac{\alpha_1}{2}a_1 + j\frac{\gamma}{2}a_2^* = 0$$

$$j\frac{\gamma}{2}a_1 - \frac{\alpha_2}{2}a_2^* = 0$$

There exists a nontrivial solution, if $\gamma^2 = \alpha_1\alpha_2$: The gain must at least equal the losses. For the intensity of the pump wave, we then get

$$I_3 = \frac{n_3}{2Z_0}|\bar{E}(0, f_3)|^2 = \frac{n_3}{2Z_0}\frac{\omega_3}{n_3}|a_3(0)|^2 = \frac{c^2}{2Z_0}\frac{n_1 n_2 n_3}{\omega_1 \omega_2 |\chi|^2}\alpha_1 \alpha_2$$

and this coincides with Eq. (5.24) since $\alpha_i' = \delta_i/L$. ∎

Frequency stability Assume that the oscillator with a resonator length L operates originally at the resonance frequencies

$$f_1 = q_1 \frac{c}{2Ln(f_1)}, \qquad f_2 = q_2 \frac{c}{2Ln(f_2)}$$

with $f_1 + f_2 = f_3$ (q_i = integer). For a small change ΔL of the resonator length due to an external disturbance, the signal and idler frequency must be shifted to the new values

$$f_1' = (q_1 + \Delta q)\frac{c}{2(L + \Delta L)n(f_1)}, \qquad f_2' = (q_2 - \Delta q)\frac{c}{2(L + \Delta L)n(f_2)}$$

with $n(f_i') \approx n(f_i) \equiv n_i$. From the requirement that $f_1' + f_2' = f_3$ still holds, we get

$$\Delta q = 2f_3 \cdot \frac{\Delta L}{L} \cdot \frac{n_1 L}{c} \cdot \frac{n_2}{n_2 - n_1}$$

We assume small values of $\Delta L/L$ and $n_2 - n_1 \ll 1$, then we find as frequency shift:

$$\delta f = f_1' - f_1 \approx \frac{c}{2Ln_1}\left[\left(q_1 + \frac{2f_3 \cdot \Delta L/L}{\frac{c}{n_1 L}\frac{n_2 - n_1}{n_2}}\right) \cdot \left(1 - \frac{\Delta L}{L}\right) - q_1\right]$$

$$\approx \frac{\Delta L}{L}\left(\frac{f_3 n_2}{n_2 - n_1} - f_1\right) \approx \frac{\Delta L}{L} \cdot \frac{f_3 n_2}{n_2 - n_1} \approx f_2 - f_2'$$

or

$$\frac{\delta f}{f_3} = \frac{\Delta L}{L} \cdot \frac{n_2}{n_2 - n_1} \tag{5.25}$$

On the other hand, a single resonator line in an SRO will be shifted by

$$\frac{\Delta f_1}{f_1} = \frac{\Delta L}{L}$$

for a fixed value of q_1 (here without the resonance condition for f_2). A change of $\Delta L/L = 10^{-7}$ thus produces a relative shift of the individual resonator frequencies also by 10^{-7}. However, the relative shift $\delta f / f_3$ of the output frequencies of the DRO, for example, increases by more than 10^{-5}, if we assume $|(n_2 - n_1)/n_2| \leq 10^{-2}$. This small frequency stability is a serious disadvantage of the DRO.

For a SRO, the frequency stability is much greater since the idler frequency can change freely according to $f_2 = f_3 - f_1$ if the resonance frequency is changed, $f_1 \rightarrow f_1'$. There is now no resonance condition, $f_2 \neq qc/(2Ln_2)$.

Tuning The parametric oscillator can be tuned by a careful variation of some tuning parameter A (for example, by a change of the refractive index—due to a change of temperature, of angle θ between the direction of incidence and the crystal axis, or by applying an electric field—or by a variation of the angles between the three waves) over a wide frequency range. From

$$\Delta k \equiv k(f_3) - k(f_1) - k(f_2) = 0, \qquad f_3 = f_1 + f_2$$

it follows that

$$\frac{\partial k(f_3)}{\partial A} - \frac{\partial k(f_1)}{\partial A} - \frac{\partial k(f_2)}{\partial A} + \frac{\partial k(f_3)}{\partial f_3}\frac{df_3}{dA} - \frac{\partial k(f_1)}{\partial f_1}\frac{df_1}{dA} - \frac{\partial k(f_2)}{\partial f_2}\frac{df_2}{dA} = 0$$

$$df_1 + df_2 = df_3 \tag{5.26}$$

For a constant pump frequency ($df_3 = 0$, $df_1 = -df_2$), we get from this at the operating point that must not coincide with phase matching ($\Delta k = $ const is sufficient, independent of A):

$$\frac{df_1}{dA} = \frac{\dfrac{\partial}{\partial A}[k(f_3) - k(f_1) - k(f_2)]}{\dfrac{\partial k(f_1)}{\partial f_1} - \dfrac{\partial k(f_2)}{\partial f_2}} \tag{5.27}$$

The measured curve in Fig. 5.3 shows the tunability of the frequencies f_1, f_2, depending on the angle θ between the resonator axis and the crystal axis; see Fig. 5.2 for an ADP crystal. As ordinate, we have plotted the angle θ' that is shifted against θ and equals zero for the case of degeneracy ($\omega_1 = \omega_3/2$). The tuning range covers almost the whole visible spectrum.

One can detune signal and idler frequency also by a direct variation of the pump frequency if one has phase matching ($\Delta k = 0$). Then the terms $\partial k/\partial A$ in Eq. (5.26) cancel, and A can be replaced by ω_3:

$$\frac{df_1}{df_3} = \frac{\dfrac{\partial k(f_3)}{\partial f_3} - \dfrac{\partial k(f_2)}{\partial f_2}}{\dfrac{\partial k(f_1)}{\partial f_1} - \dfrac{\partial k(f_2)}{\partial f_2}} \tag{5.28}$$

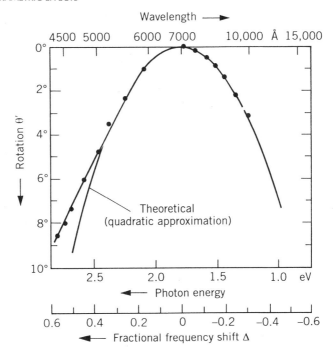

FIGURE 5.3 Tuning behavior for the parametric oscillator. (Reproduced with permission of D. Magde, from *Phys. Rev. Lett.* 18 (1967): 905.)

PROBLEMS

Section 5.1

5.1 Describe a children's swing as a degenerate parametric oscillator $f_3 = f_1 + f_1$.

5.2 Derive Eq. (5.6).

Section 5.3

5.3 Derive Eq. (5.13).

5.4 Calculate the upconversion efficiency for the following process: A pump wave from an Nd:YAG laser ($\lambda_2 = 1.06$ μm) is mixed with a weak signal from a CO_2 laser ($\lambda_1 = 10.6$ μm), yielding the upconverted wave ($\lambda_3 = 0.963$ μm). The intensity of the pump wave is 10^8 W/m^2. As nonlinear crystal, we use proustite with $L = 1$ cm, $n(f_1) \approx n(f_2) \approx n(f_3) = 2.6$, and $\chi = 3.4 \times 10^{-11}$ m/V.

5.5 Derive the expression for down conversion efficiency, $\eta = \langle S(L, f_1) \rangle / \langle S(0, f_3) \rangle$.

Section 5.4

5.6 A LiNbO$_3$ crystal ($\chi = 0.56 \times 10^{-11}$ m/V) is pumped with an (unrealistic) high pump intensity of $I = 5 \times 10^6$ W $\cdot$ cm^{-2}. We have $f_3 = f_1 + f_1$, $f_1 = 3 \times 10^{14}$ Hz, and $n_1 = n_3 = 2.2$. Calculate the gain constant κ.

5.7 Derive Eq. (5.25).

REFERENCES

D. Magde. *Phys. Rev. Lett. 18* (1967). 905.

BIBLIOGRAPHY

G. P. Agrawal. *Nonlinear Fiber Optics*. Wiley, New York, 1984.

R. W. Boyd. *Nonlinear Optics*. Academic Press, San Diego, CA, 1992.

G. K. Grau. *Quantenelektronik*. Vieweg, Braunschweig, 1978.

B. E. A. Saleh and M. C. Teich. *Fundamentals of Photonics*. Wiley, New York, 1991.

Y. R. Shen. *The Principles of Nonlinear Optics*. Wiley, New York, 1984.

A. Yariv. *Quantum Electronics*. Wiley, New York, 1975.

A. Yariv and J. E. Pearson. *Parametric processes*. In *Progress in Quantum Electronics* (J. H. Sanders and K. W. Stevens, eds.), Vol. 1, Part 1. Pergamon Press, Oxford, 1969.

CHAPTER SIX

Raman and Brillouin Effect

The wave coupling phenomena that we are discussing are not restricted only to electromagnetic waves. For example, one of the participating electromagnetic waves can be replaced by a material excitation (elastic vibrations in crystals or molecular vibrations); this is the case for the Raman and Brillouin effect.

6.1 QUALITATIVE CONSIDERATIONS ON THE RAMAN EFFECT

In inelastic scattering of light from molecules, the *(spontaneous) Raman effect* (SRE) occurs, and this happens at relatively low excitation energies: A pump wave $\mathbf{E} = \mathbf{E}_p \cdot \sin \omega_p t$ is incident on a molecule. Under the influence of the field, the electrons of the molecule oscillate against the heavy nuclei, and an electric dipole moment $\mathbf{m}$ arises proportional to $\mathbf{E}_p$ for small field amplitudes:

$$\mathbf{m} = \alpha \cdot \mathbf{E} = \alpha \cdot \mathbf{E}_p \cdot \sin \omega_p t \tag{6.1}$$

α is the polarizability of the molecule. If there are now, in addition, internal molecular vibrations with a frequency ω_m, for example, due to oscillations of some nuclei against one another, the polarizability can change periodically with ω_m:

$$\alpha = \alpha_0 + \alpha_1 \cdot \sin \omega_m t \tag{6.2}$$

Thus, from Eq. (6.1) follows:

$$\mathbf{m} = \alpha_0 \mathbf{E}_p \cdot \sin \omega_p t + \left(\frac{\alpha_1 \mathbf{E}_p}{2} \right) [\cos(\omega_p - \omega_m)t - \cos(\omega_p + \omega_m)t] \tag{6.3}$$

Therefore, the oscillating dipole moment $\mathbf{m}$ of the molecule contains three frequencies that are emitted. To begin with, the frequency ω_p of the incoming

77

light appears again: This is *elastic Rayleigh scattering*. But there are still two new frequencies:

$$\omega_s = \omega_p - \omega_m, \qquad \omega_{as} = \omega_p + \omega_m, \tag{6.4}$$

referred to as the *Stokes* and the *anti-Stokes frequencies*. Equation (6.4) represents, after multiplication with $\hbar$, conservation of energy for the Stokes and anti-Stokes process:

$$\hbar\omega_p = \hbar\omega_s + \hbar\omega_m, \qquad \hbar\omega_{as} = \hbar\omega_p + \hbar\omega_m \tag{6.5}$$

As a dynamical model for the Raman effect, we consider vibrational oscillations in the molecule, analogously to the treatment of the anharmonic oscillator discussed in Section 1.4. The Raman medium consists of independent oscillators performing forced and damped oscillations. For the deviation X from the equilibrium position of one of these oscillators (at position z), we have as the equation of motion

$$\frac{d^2 X(z,t)}{dt^2} + \gamma \frac{dX}{dt} + \omega_m^2 X = \frac{F(z,t)}{m_0} \tag{6.6}$$

γ is a damping constant, ω_m a resonance frequency of the molecule. The driving force $F(z,t)$ can be calculated from the density of the stored electrical energy [Eq. (2.38) in the isotropic case]:

$$w_{el} = \left(\frac{\varepsilon}{2}\right) \cdot E^2$$

ε can be derived microscopically from the spatially dependent polarizability α and so is itself dependent on X:

$$\varepsilon(X) = \varepsilon_0 + \left(\frac{\partial\varepsilon}{\partial X}\right)_0 \cdot X + \dots$$

The applied optical field exerts the force

$$F(z,t) = \frac{dw_{el}}{dX} = \frac{1}{2}\frac{\partial\varepsilon}{\partial X}\langle E^2(z,t)\rangle \approx \frac{1}{2}\left(\frac{\partial\varepsilon}{\partial X}\right)_0 \langle E^2(z,t)\rangle = C_1\langle E^2(z,t)\rangle$$

$\langle\,\rangle$ stands for temporal averaging over few optical periods. This averaging is necessary because the molecular oscillations cannot keep step with the optical oscillations. For the electrical field, we take the sum of a laser pump field (f_p) and Stokes field (f_s):

$$E(z,t) = \frac{1}{2}(\hat{E}_p e^{j\omega_p t} + \hat{E}_s e^{j\omega_s t} + \text{c.c.}) \tag{6.7}$$

Thus, with the abbreviation

$$\Omega = \omega_p - \omega_s$$

the temporally variable part of $F(z, t)$ is

$$F(z, t) = \frac{C_1}{2} [\hat{E}_p(z) \hat{E}_s^*(z) e^{j\Omega t} + \text{c.c.}]$$

since the parts with $2\omega_p, 2\omega_s, \omega_p + \omega_s$ cancel if averaged over time. The ansatz

$$X(z, t) = \frac{1}{2} [\hat{X}(z) e^{j\Omega t} + \text{c.c.}] \tag{6.8}$$

yields using the differential equation (6.6) immediately

$$(-\Omega^2 + j\gamma\Omega + \omega_m^2)\hat{X} = \left(\frac{C_1}{m_0}\right) \cdot \hat{E}_p \hat{E}_s^*$$

or

$$\hat{X}(z) = \frac{C_1}{m_0} \frac{\hat{E}_p(z)\hat{E}_s^*(z)}{\omega_m^2 - \Omega^2 + j\gamma\Omega} \tag{6.9}$$

The polarization $P(z, t)$ of the medium can be expressed with the induced dipole moment, Eq. (6.1), $m(z, t) = \alpha(z, t)E(z, t)$ in the form

$$P(z, t) = Nm(z, t) = N\alpha(z, t)E(z, t)$$
$$= N[\alpha_0 + \left(\frac{\partial\alpha}{\partial X}\right)_0 X + \ldots] \cdot E(z, t) = P_L + P_{NL}$$

with N as the number of oscillators per unit volume. Substitution of X according to Eqs. (6.8) and (6.9) and of E according to Eq. (6.7) yields as the nonlinear part of polarization

$$P_{NL} = N\left(\frac{\partial\alpha}{\partial X}\right)_0 \frac{1}{2} \left[\frac{C_1}{m_0} \frac{\hat{E}_p(z)\hat{E}_s^*(z)}{\omega_m^2 - \Omega^2 + j\gamma\Omega} e^{j\Omega t} + \text{c.c.}\right] \frac{1}{2} \left(\hat{E}_p e^{j\omega_p t} + \hat{E}_s e^{j\omega_s t} + \text{c.c.}\right)$$

This expression contains many different frequency contributions. We are only interested in that part which oscillates with the Stokes frequency $\omega_s = \omega_p - \Omega$:

$$P_{NL}(z, t; \omega_s) = \frac{C_2'}{2} \left[\frac{\hat{E}_p^*(z)\hat{E}_s(z)}{\omega_m^2 - \Omega^2 - j\gamma\Omega} \hat{E}_p e^{-j\Omega t + j\omega_p t} + \text{c.c.}\right]$$
$$= \frac{1}{2} [\hat{P}_{NL}(z, t; \omega_s) e^{j\omega_s t} + \hat{P}_{NL}^*(z, t; \omega_s) e^{-j\omega_s t}]$$

Thus,

$$\hat{P}_{NL}(z,t;\omega_s) = C_2' \frac{|\hat{E}_p(z)|^2 \hat{E}_s(z)}{\omega_m^2 - \Omega^2 - j\gamma\Omega} = \frac{C_2' \hat{E}_p(z)\hat{E}_p^*(z)\hat{E}_s(z)}{\omega_m^2 - (\omega_p - \omega_s)^2 - j\gamma(\omega_p - \omega_s)} \quad (6.10)$$

We therefore have with Raman scattering the first case of a four-wave mixing process, that is, three fields interact and produce a fourth field.

Explanation in a corpuscular picture   See Fig. 6.1. A photon with energy $\hbar\omega_p$ hits a molecule that is in the ground state (lowest energy level, $v = 0$) of molecular oscillations:

$$W_v = \hbar\omega_m\left(v + \frac{1}{2}\right), \qquad v = 0, 1, 2, \dots \quad (6.11)$$

The molecule absorbs the photon temporarily and is then in a forbidden (virtual) energy state, since in general $W_0 + \hbar\omega_p$ does not coincide with an allowed energy level W_v. From this virtual energy level, the molecule will return very soon in an energy state W_v. There are two possibilities then:

1. The molecule returns into the ground state ($v = 0$) and the whole energy $\hbar\omega_p$ is emitted again. The photon has only changed its direction (arbitrarily), but has transferred no energy to the molecule. This is elastic Rayleigh scattering, a parametric process.
2. The molecule returns to the first excited state ($v = 1$). Here, a photon with the energy $\hbar\omega_s = \hbar\omega_p - \hbar\omega_m$ is emitted and all directions of emissions have equal probability. This is the spontaneous Raman scattering in the Stokes case.

However, if initially the molecule is already in the first excited state W_1, then besides Rayleigh scattering and Stokes scattering into higher states, the possibility exists that the molecule which has been brought into a virtual state by the absorption of $\hbar\omega_p$ returns to the ground state W_0. Here, a photon with energy $\hbar\omega_{as} = \hbar\omega_p + \hbar\omega_m$ must be emitted: This is the spontaneous Raman scattering in the anti-Stokes case. Whereas the Stokes radiation lies on the long wavelength

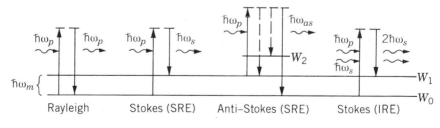

FIGURE 6.1   Raman effect.

side of the incident spectral line, the anti-Stokes radiation is shifted to shorter wavelengths. Moreover, there are always considerably less molecules in state W_1 (i.e., N_1) than state W_0 (number N_0) in thermal equilibrium at an absolute temperature T according to the Boltzmann distribution

$$N_1 = N_0 \cdot \exp\left(\frac{-\hbar\omega_m}{k_B T}\right) \tag{6.12}$$

Thus, the anti-Stokes lines are always much weaker than the Stokes lines at thermal equilibrium. The frequency shift given by ω_m varies, of course, for different Raman materials and lies between about 100 cm^{-1} and some few 10^3 cm^{-1}; the relative frequency shift is $\Delta f/f \approx 10^{-3} \ldots 10^{-2}$.

For large light intensities (laser light), besides increased spontaneous Raman scattering, however, another effect occurs: the *"induced"* or *"stimulated"* *Raman effect* (IRE). An incoming photon $\hbar\omega_p$ gives rise to the transition of the molecule from the ground state to a virtual level, where it remains for a short duration. If now within this brief time, a Stokes photon $\hbar\omega_s$ that has been created elsewhere hits this molecule, a so-called induced emission can result, that is, a Stokes photon $\hbar\omega_s$ will be created, which coincides exactly with the incoming Stokes photon in direction, frequency, phase, and polarization. In contrast to the incoherent SRE, we have here coherent radiation; there is a fixed phase relation between the exciting wave and generated Stokes wave. As energy balance, we simply have:

$$\hbar\omega_p + \hbar\omega_s = 2\hbar\omega_s + \hbar\omega_m \tag{6.13}$$

In experiments, one finds that Stokes and anti-Stokes waves can be generated simultaneously. This takes place as follows: After a photon with energy $\hbar\omega_p$ has created a Stokes photon, the molecule is in the excited state W_1. This energy is immediately used by a second photon with $\hbar\omega_p$ for the emission of an anti-Stokes photon. The molecule does not change its state; we have then a parametric process: *Stokes–anti-Stokes coupling*. Now, Stokes and anti-Stokes waves with comparable intensities are found. The energy balance is

$$2\hbar\omega_p = \hbar\omega_s + \hbar\omega_{as} \tag{6.14}$$

In parametric effects, a momentum balance must be fulfilled among the propagation vectors of the participating light waves:

$$2\mathbf{k}_p = \mathbf{k}_s + \mathbf{k}_{as} \tag{6.15}$$

with magnitudes

$$|\mathbf{k}_p| = \frac{\omega_p n_p}{c}, \qquad |\mathbf{k}_s| = \frac{\omega_s n_s}{c}, \qquad |\mathbf{k}_{as}| = \frac{\omega_{as} n_{as}}{c}$$

Because the frequencies lie very close together, one can expand in the neighborhood of $\omega = \omega_p$:

$$n_s = n(\omega_s) = n_p - \omega_m \cdot \frac{dn}{d\omega}, \qquad n_{as} = n(\omega_{as}) = n_p + \omega_m \cdot \frac{dn}{d\omega}$$

where $\omega_m = \omega_p - \omega_s = \omega_{as} - \omega_p$. Thus,

$$|\mathbf{k}_s| + |\mathbf{k}_{as}| = 2|\mathbf{k}_p| + \frac{2}{c} \cdot \omega_m^2 \cdot \frac{dn}{d\omega} \tag{6.16}$$

In the region of normal dispersion $dn/d\omega > 0$, so

$$|\mathbf{k}_s| + |\mathbf{k}_{as}| > 2|\mathbf{k}_p|$$

that is, the required conservation of momentum, Eq. (6.15), cannot be satisfied with parallel propagation vectors; Eq. (6.15) constitutes a nondegenerate triangle. Stokes and anti-Stokes light is radiated in the form of a conical shell about the direction of incidence with a definite aperture angle. In contrast, the SRE yields a nearly isotropic emission.

In the Raman effect with molecular vibrations, a localized molecule takes part changing its state. This is the reason why no phase matching is necessary here.

In the Raman effect within a crystal, an interaction with the elastic lattice vibrations occurs: These lattice vibrations (phonons from the optical branch of the dispersion diagram, see Fig. 6.2) are not localized; they belong to the whole crystal and behave as if they had a definite energy ($\hbar\omega_m$) and definite momentum ($\hbar k_m$). Conservation of energy and momentum must be fulfilled:

$$\omega_p = \omega_s + \omega_m, \qquad \mathbf{k}_p = \mathbf{k}_s + \mathbf{k}_m \tag{6.17}$$

If, in particular, all vectors $\mathbf{k}$ are parallel, $k_p = k_s \pm k_m$, then we find from

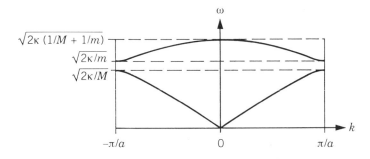

FIGURE 6.2 Dispersion relation of a diatomic chain of equidistantly lying masses $M > m$ with the optical (= higher) and the acoustical (= lower) branch. κ = coupling constant. Distance between two neighbouring masses M, m is $a/2$.

Eq. (6.17)

$$\omega_m = \omega_p \left[1 - \frac{n(f_p)}{n(f_s)} \right] \pm \frac{ck_m}{n(f_s)}$$

By an appropriate choice of $n(f_p)/n(f_s)$ in a birefringent crystal (one wave as an ordinary wave, the other as extraordinary), the value of ω_m and thus ω_s can be changed. We obtain a Stokes wave that has a continuously variable frequency and can be amplified by induced scattering. This represents one method for the generation of coherent, far-infrared radiation.

The transition rate (probability per unit time) for spontaneous emission of Stokes photons is (this is a result of a quantum theoretical calculation) given by

$$P_{se}(\hbar\omega_s) = D \cdot q_p \cdot N_0 \qquad (6.18)$$

where q_p is the number of incident pump photons with energy $\hbar\omega_p$, D a constant of proportionality. The transition rate for the spontaneous emission of anti-Stokes photons is correspondingly

$$P_{se}(\hbar\omega_{as}) = D \cdot q_p \cdot N_1 \qquad (6.19)$$

The probability for induced emission is, however, always proportional to the number q_s of initially existing Stokes photons, so the transition rate for the induced emission of Stokes photons is

$$P_{ie}(\hbar\omega_s) = D \cdot q_p \cdot q_s \cdot N_0 \qquad (6.20)$$

Here, a product of two photon numbers appears: This is a process nonlinear in the optical fields. The transition rate into the Stokes mode is thus altogether

$$P(\hbar\omega_s) = Dq_p N_0(q_s + 1) = \frac{dq_s}{dt} \qquad (6.21)$$

and equals the rate of change for the mean photon number q_s in Stokes light. Then, the spatial change on a distance $z = vt = (c/n)t$ is given by

$$\frac{dq_s}{dz} = \frac{n}{c} \frac{dq_s}{dt} = \frac{n}{c} \cdot Dq_p N_0(q_s + 1) \qquad (6.22)$$

The case $q_s \ll 1$ corresponds again to SRE. Integration yields here (for $q_p = \text{const}$, i.e., no pump depletion)

$$q_s(z) = \frac{n}{c} \cdot Dq_p N_0 \cdot z \qquad (6.23)$$

This results in the linear growth of the Stokes intensity $q_s \cdot \hbar\omega_s$ with length and proportional to the number N_0 of molecules of the ground state in the interaction region. The case $q_s \gg 1$ corresponds to IRE. Integration yields now

$$q_s(z) = q_s(0) \cdot \exp(Gz) \tag{6.24}$$

with the Raman gain coefficient $G = (n/c)DN_0 q_p$. Due to this exponential increase, the IRE is usually considerably stronger than the SRE: In the SRE, only a fraction of 10^{-6} of the incident radiation is scattered after penetrating 1 cm of the scattering medium; on the other hand, with the IRE, 10% or more is scattered. In the SRE, the spectral width of the output radiation corresponds to the width of the vibrational level of the medium. On the other hand, in IRE the stimulated output occurs mainly at that frequency where the gain is highest. As a result, the spectral width of the output is narrow, compared with that of the spontaneous emission.

Raman spectroscopy is used as a mean for the investigation of just the vibrational levels of the molecules or lattice vibrations in the crystal. If the ground state $|W_0\rangle$ and final state $|W_1\rangle$ possess the same parity, an electric dipole transition $\langle W_1|\hat{\mu}|W_0\rangle$ is forbidden in materials with inversion symmetry ($\hat{\mu}$ is the operator of the electric dipole moment). However, with the Raman effect, this very transition can be investigated if the intermediate state possesses the other parity.

Example 6.1   Calculate the frequency of the natural vibrations of a bromine molecule, if in Raman scattering, a pump line with $\lambda_p = 313.16$ nm produces a Stokes line with $\lambda_s = 316.4$ nm. ($f_m = 10^{13}$ Hz.) ∎

6.2   INDUCED RAMAN SCATTERING

If an intense light beam enters, for example, a cuvette filled with a liquid, at first spontaneously generated Stokes photons $\hbar\omega_s$ arise. Some of them run in the direction of the original light beam and can create still more Stokes photons by induced emission. Those Stokes photons that run parallel to the incident light beam are amplified; the intensity of the Stokes photons increases at the expense of the incident pump photons $\hbar\omega_p$, whose number decreases with the increasing length of the cuvette. The spontaneous effect can only be described by investigating the pertinent quantum dynamics.

Thus, in the following we consider only induced processes. In the external electric field, only the frequencies ω_p and ω_r ($=\omega_s$, or $=\omega_{as}$) occur:

$$\mathbf{E}(\mathbf{x}, t) = \mathbf{E}_p(\mathbf{x}, t) + \mathbf{E}_r(\mathbf{x}, t) \tag{6.25}$$

Therefore, a nonlinear polarization at the frequencies ω_p and ω_r (in lowest order) can only occur via the processes

$$\omega_p = \omega_r - \omega_r + \omega_p, \qquad \text{or} \qquad \omega_r = \omega_p - \omega_p + \omega_r \tag{6.26}$$

Vibrations of the material are contained in the susceptibility.

Therefore, in a *classical electro-dynamical treatment,* one has to use third-order nonlinear polarization at a laser frequency f_p and a Raman frequency f_r of the form [See Eq. (1.59)]:

$$\hat{P}_i^{(3)}(f_p) = \left(\frac{3\varepsilon_0}{2}\right) \sum \chi_{ijkl}(-f_p, f_r, -f_r, f_p)\hat{E}_j(f_r)\hat{E}_k^*(f_r)\hat{E}_l(f_p)$$

$$\hat{P}_i^{(3)}(f_r) = \left(\frac{3\varepsilon_0}{2}\right) \sum \chi_{ijkl}(-f_r, f_p, -f_p, f_r)\hat{E}_j(f_p)\hat{E}_k^*(f_p)\hat{E}_l(f_r)$$

(6.27)

This is now the first example of four-wave mixing (FWM).

In a *semi-classical description* (only the material excitations are quantized) using the density matrix formalism, a third-order perturbation theory is required since the nonlinear polarizations, Eq. (6.27), are proportional to the third power of the components of the electric field.

Finally, in a *fully quantized theory* (here, both the light waves and material excitations are quantized), the Raman effect has to be considered a two-photon process (one photon ω_p is absorbed and another photon ω_r emitted), while simultaneously, the material undergoes a transition from an initial state to a final one, that is, one has to use second-order perturbation theory according to Dirac. But both quantum theoretical treatments lie beyond the scope of this book.

In the IRE, we can thus use classically a coupling of a pump and a Raman wave mediated by a third-order process. We treat the IRE with molecular vibrations. Comparing Eq. (6.27) with Eq. (6.10), one finds that $\chi_{ijkl}(-f_r, f_r, f_p, -f_p)$ has the structure

$$\chi_{iikl}(-f_r, f_r, f_p, -f_p) = \frac{C_2}{\omega_m^2 - (\omega_p - \omega_s)^2 - j\gamma(\omega_p - \omega_s)}$$

(6.28)

It follows from this that χ_{ijkl} is imaginary for $\omega_s = \omega_p - \omega_m$ and $\omega_{as} = \omega_p + \omega_m$ (see Problem 6.1) with

$$\chi_{ijkl} = \pm j|\chi_{ijkl}|, \quad \text{Im } \chi_{ijkl} = \pm|\chi_{ijkl}| \quad \text{for} \quad f_r = \begin{cases} f_s \\ f_{as} \end{cases}$$

in contrast to the ideal parametric processes, where χ is real. We introduce again effective nonlinear coefficients:

$$\chi_{\text{eff}}(-f_p, f_r, -f_r, f_p) = \frac{3}{2}\sum \chi_{ijkl}(-f_p, f_r, -f_r, f_p)e_i(f_p)e_j(f_r)e_k(f_r)e_l(f_p)$$

$$\chi_{\text{eff}}(-f_r, f_p, -f_p, f_r) = \frac{3}{2}\sum \chi_{ijkl}(-f_r, f_p, -f_p, f_r)e_i(f_r)e_j(f_p)e_k(f_p)e_l(f_r)$$

(6.29)

with the symmetry relation, following from Eq. (1.33), Eq. (1.38):

$$\chi_{\text{eff}}^*(-f_p, f_r, -f_r, f_p) = \chi_{\text{eff}}(-f_r, f_p, -f_p, f_r) =: \chi \qquad (6.30)$$

For χ holds exactly the same as for χ_{ijkl}:

$$\chi = \pm j \cdot |\chi|, \quad \text{Im } \chi = \pm|\chi| \quad \text{for} \quad f_r = \begin{cases} f_s \\ f_{as} \end{cases} \qquad (6.31)$$

In the approximation of the slowly varying amplitude, Eq. (2.21) has to be written down for the pump wave (f_p) and the Raman wave (f_r):

$$\frac{\partial}{\partial z}\bar{E}(z, f_p) = -j\frac{\omega_p^2 \mu_0}{2k(f_p)} \cdot \sum \hat{P}_i(f_p)e_i(f_p)e^{jk(f_p)z} = -j\frac{\omega_p}{2n_p c} \cdot \chi^*|\bar{E}(f_r)|^2\bar{E}(f_p)$$

$$\frac{\partial}{\partial z}\bar{E}(z, f_r) = -j\frac{\omega_r^2 \mu_0}{2k(f_r)} \cdot \sum \hat{P}_i(f_r)e_i(f_r)e^{jk(f_r)z} = -j\frac{\omega_r}{2n_r c} \cdot \chi \cdot |\bar{E}(f_p)|^2\bar{E}(f_r)$$

$$(6.32)$$

Since $\exp\{jz[k(f_p) - k(f_r) + k(f_r) - k(f_p)]\} = 1$, the oscillating phase factor on the right-hand side is cancelled; as already mentioned, no particular phase matching is necessary. We introduce the abbreviations

$$g_r = \frac{\omega_r}{n_r c} \text{ Im } \chi, \qquad g_r' = g_r \frac{2Z_0}{n_p} \qquad (6.33)$$

Then, from Eq. (6.32) with Eq. (A.6), we get for the intensities

$$\frac{\partial}{\partial z}I(z, f_p) = -\frac{\omega_p}{\omega_r}g_r'I(z, f_p)I(z, f_r)$$

$$\frac{\partial}{\partial z}I(z, f_r) = g_r'I(z, f_p)I(z, f_r) \qquad (6.34)$$

Moreover, we see easily that

$$\frac{I(z, f_p)}{\omega_p} + \frac{I(z, f_r)}{\omega_r} = \frac{I(0, f_p)}{\omega_p} + \frac{I(0, f_r)}{\omega_r} = \text{const} = K > 0 \qquad (6.35)$$

is constant, independently of z. This means simply that the total photon number in the pump and the Stokes wave (or anti-Stokes wave) is constant: For each annihilated pump photon, a Stokes photon is created.

According to Eq. (6.33), we need for the Raman gain spectrum ($\Omega = \omega_p - \omega_s$)

$$\operatorname{Im} \chi = \operatorname{Im}\ \frac{C_2}{\omega_m^2 - \Omega^2 - j\gamma\Omega} = \operatorname{Im}\ \frac{C_2}{(\omega_m - \Omega)(\omega_m + \Omega) - j\gamma\Omega}$$

In order to determine the gain spectrum, we do not work directly at resonance $\omega_m = \Omega$, but at least close to it: $\omega_m \approx \Omega$. Then we get

$$\operatorname{Im} \chi \approx \operatorname{Im} \frac{C_2}{2\omega_m(\omega_m - \Omega - j\gamma/2)} = \frac{C_2\gamma/2}{2\omega_m\{[\omega_m - (\omega_p - \omega_s)]^2 + \gamma^2/4\}}$$

Thus,

$$g_r = \frac{\omega_r C_2}{n_r c \gamma \omega_m} \cdot \frac{1}{1 + \left[\dfrac{\omega_m - (\omega_p - \omega_s)}{\gamma/2}\right]^2} \tag{6.36}$$

and this is a Lorentz distribution with a FWHM $\gamma \equiv \Delta f_r$ as the Raman-linewidth. A typical value is

$$\gamma \lesssim 10^{-2}\omega_m = 2\pi \times 0.3 \times 10^{12}\mathrm{s}^{-1} = 2\pi \times 0.3\,\mathrm{THz}$$

for $\lambda_m = (1/992)$ cm $= 10.081\ \mu$m.

For strong pump radiation, we take again $I(z, f_p) \approx I(0, f_p) = $ const; from the second Eq. (6.34), it then follows that

$$I(z, f_r) = I(0, f_r) \cdot \exp[g_r' \cdot I(0, f_p) \cdot z] = I(0, f_r) \cdot \exp[g_r \cdot |\bar{E}(0, f_p)|^2 \cdot z] \tag{6.37}$$

g_r or g_r' are the *Raman gain coefficients*. Thus, we find an exponential increase of $I(z, f_r)$ at the Stokes frequency f_s and a decrease of $I(z, f_r)$ at the anti-Stokes frequency f_{as} (since here $g_r < 0$).

Example 6.2 For benzene, $g_r' = 2.8 \times 10^{-3}$ cm/MW at $\lambda = (1/992)$ cm. If we ask for a gain of $\mathcal{P}(z, f_r)/\mathcal{P}(0, f_r) = I(z, f_r)/I(0, f_r) = e^{30} = 1.07 \times 10^{13}$, we can obtain this in a 10-cm-long benzene cell with $I(0, f_p) \approx 1$ GW/cm^2. In fact, 100 MW/cm^2 or still less are already sufficient. The reason for this reduction of pump intensity is the self-focusing of the laser beam in the nonlinear medium (see Section 7.3). ∎

However, Eqs. (6.34) can be solved also in the case of pump depletion using Eq. (6.35). We find

$$I(z, f_p) = \frac{K\omega_p}{1 + \dfrac{\omega_p}{\omega_r}\dfrac{I_{r0}}{I_{p0}} \cdot \exp(g_r'\omega_p K z)}$$

$$I(z, f_r) = \frac{K\omega_r}{1 + \dfrac{\omega_r}{\omega_p} \dfrac{I_{p0}}{I_{r0}} \cdot \exp(-g'_r \omega_p K z)} \tag{6.38}$$

The Stokes wave ($r = s$, $g'_r > 0$) will be amplified, the anti-Stokes wave ($r = as$, $g'_r < 0$) deamplified. An amplification of the anti-Stokes wave is possible only in parametric four-wave processes. For large values of z, the intensities $I(z, f_p) \to 0$ and $I(z, f_s) \to K\omega_s$ in the Stokes case, whereas in the anti-Stokes case, $I(z, f_p) \to K\omega_p$ and $I(z, f_{as}) \to 0$.

For the Raman oscillator, the Raman-active medium is put into an optical resonator that is resonant at the Stokes frequency. As soon as the gain compensates the losses in a whole round-trip, oscillations can start. Raman oscillation is used to convert an original frequency of pulsed lasers into coherent light at the Raman-shifted frequencies.

6.3 BRILLOUIN EFFECT

In contrast to the Raman effect where optical phonons arise, acoustical phonons participate in the Brillouin effect; see Fig. 6.2. The induced Brillouin effect (IBE) can be described classically as a parametric process between a pump wave ($\omega_p, \mathbf{k}_p$), a second optical wave ($\omega_b, \mathbf{k}_b$), and an acoustical wave ($\omega_a, \mathbf{k}_a$). The pump wave generates via the process of electrostriction in a nonlinear medium (liquid or crystal) an acoustical wave (compression wave), which causes a periodical modulation of the refractive index. The pump wave is Bragg-reflected from this pump-induced grating. The scattered light has a new frequency (ω_b) because of the Doppler shift at the grating that is moving with the acoustical velocity v_a.

The quantum theoretical explanation describes the process as the annihilation of a pump photon ω_p, with the simultaneous creation of a new optical photon (ω_b) and an acoustical phonon (ω_a). These acoustical phonons belong to the whole lattice; they are not localized and have a definite energy and definite momentum. In the scattering process, energy and momentum are conserved between both photons and the phonon:

$$\omega_b = \omega_p \mp \omega_a, \qquad \mathbf{k}_b = \mathbf{k}_p \mp \mathbf{k}_a \quad \begin{cases} \text{Stokes case} \\ \text{anti-Stokes case} \end{cases} \tag{6.39}$$

See Eq. (6.4).

Due to thermal motion of the lattice constituents (or molecules), acoustical vibrations are already excited at normal temperatures. Therefore, in the Brillouin effect, Stokes and anti-Stokes frequencies appear with comparable intensities. Let

$$v_a = \frac{\omega_a}{k_a}, \qquad v_p = \frac{c}{n_p} = \frac{\omega_p}{k_p}$$

be the phase velocities of the acoustical wave and the incident light wave, respectively, with a ratio of $v_a/v_p \sim 10^{-5}$. If for the magnitudes of the phonon and pump photon wave vectors, $|\mathbf{k}_a| \leq |\mathbf{k}_p|$ holds, there follows $\omega_a = v_a k_a \ll v_p k_p = \omega_p$. Thus, $\omega_p \approx \omega_b, n_p \approx n_b, v_p \approx v_b$ and $|\mathbf{k}_p| \approx |\mathbf{k}_b|$. With γ as the angle between both light wave vectors $\mathbf{k}_p$ and $\mathbf{k}_b$, we get from Eq. (6.39)

$$\mathbf{k}_a^2 = \mathbf{k}_b^2 + \mathbf{k}_p^2 - 2|\mathbf{k}_b| \cdot |\mathbf{k}_p| \cdot \cos\gamma \approx 2\mathbf{k}_p^2(1 - \cos\gamma) = 4\mathbf{k}_p^2 \cdot \sin^2\frac{\gamma}{2} \qquad (6.40)$$

Then we obtain

$$\omega_a = 2 \cdot \frac{v_a}{v_p} \cdot \omega_p \sin\frac{\gamma}{2} = \frac{4\pi v_a n_p}{\lambda_p} \cdot \sin\frac{\gamma}{2}, \qquad \omega_b = \omega_p \mp \omega_a = \omega_p\left(1 \mp 2 \cdot \frac{v_a}{v_p} \cdot \sin\frac{\gamma}{2}\right)$$
$$(6.41)$$

In contrast to Raman scattering, the frequency shift is now dependent on angle. In the forward direction ($\gamma = 0$), there is no Brillouin scattering: $\omega_a = 0, \mathbf{k}_a = 0$, $\omega_b = \omega_p$. The largest value for the frequency shift is obtained for $\gamma = \pi$; then, the pump light wave and the frequency-shifted wave generated by the Brillouin effect run in opposite directions.

Example 6.3   The scattering of visible light with $\lambda_p = 0.4$ μm $[k_p = k(f_p) = 2\pi/\lambda_p \cdot n_p = (\pi/2) \times 1.5 \times 10^5$ cm^{-1}, at $n_p = 1.5]$, $v_a = 5 \times 10^5$ cm/s yields as the maximum phonon frequency (for scattering in the backward direction) $\omega_a \approx 2 \cdot 10^{11}$ s^{-1} and $k_a \approx 5 \cdot 10^5$ cm^{-1}. The relative frequency change $|\omega_b - \omega_p|/\omega_p = \omega_a/\omega_p$ has then a maximum with an order of magnitude of 10^{-5} and is thus much less than with the Raman effect.   ■

In general, momentum conservation, Eq. (6.39), constitutes an almost isosceles triangle for the momentum vectors; for the components of the momenta parallel to $\mathbf{k}_a$ with $\lambda_p \approx \lambda_b$ the Bragg condition ($|\mathbf{k}_a| = 2\pi/\Lambda$) follows:

$$2\Lambda \cdot \sin\frac{\gamma}{2} = \frac{\lambda_p}{n_p} \qquad (6.42)$$

The scattered (reflected) wave must superpose correctly in phase with the incident wave in order to not be extinguished by destructive interference.

The frequency shift, Eq. (6.41), can also be found by calculating the reflection of an incident wave from a moving mirror (the Doppler effect, see Problem 6.5).

The differential equations in the case of backward scattering ($\gamma = \pi$) result from those of the Raman effect, Eq. (6.34), if one takes into account the fact that for the generated wave in the backward direction, the signs on the right-hand side must be inverted. Due to the smallness of the frequency shift, we put here $\omega_p \approx \omega_b$. Thus,

$$\frac{dI_p}{dz} = -g'_b I_p I_b$$

$$\frac{dI_b}{dz} = -g'_b I_p I_b$$

(6.43)

or (conservation of energy)

$$\frac{d}{dz}(I_p - I_b) = 0, \qquad I(z, f_p) - I(z, f_b) = I_{p0} - I_{b0} = K' = \text{const} \qquad (6.44)$$

Here, g'_b is the Brillouin gain coefficient with a maximum at ω_b. For the gain spectrum, we refer to Eq. (6.36); however, it is now very small (typically 100 MHz). Equations (6.43) can be solved analytically:

$$I_p(z) = \frac{K'}{1 - \dfrac{I_{b0}}{I_{p0}} \exp(-g'_b K' z)}$$

$$I_b(z) = \frac{K'}{\dfrac{I_{p0}}{I_{b0}} \exp(g'_b K' z) - 1}$$

(6.45)

For a constant pump field I_p, however, one gets more easily from Eq. (6.43)

$$I_b(z) = I_b(L) \cdot \exp[g'_b \cdot I_p \cdot (L - z)] \qquad (6.46)$$

If no power is fed in, that is, $I_b(L) = 0$, the wave I_b develops due to spontaneous Brillouin scattering.

PROBLEMS

Section 6.1

6.1 Derive a classical expression for the nonlinear polarization in the anti-Stokes case analogously to the derivation of Eq. (6.10) and show that $\chi_{ijkl} = -j|\chi_{ijkl}|$.

6.2 At thermal equilibrium with an absolute temperature T, the Boltzmann distribution gives the relative populations of two energy levels W_1 and W_2: $N_2 = N_1 \exp[-(W_2 - W_1)/k_B T]$ with $W_2 - W_1 = hf_{21}$. Using the

Boltzmann constant $k_B = 1.38 \times 10^{-23}$ Ws/K, calculate the ratio N_2/N_1
   (a) for an optical transition $\lambda_{21} = 500$ nm at $T = 300$ K.
   (b) for a wave number $1/\lambda = 100$/cm at $T = 300$ K.
   (c) for a microwave transition $f_{21} = 3$ GHz at $T = 300$ K.
   (d) for a 10-GHz transition at liquid-helium temperature $T = 4.2$ K.

6.3 Compare Stokes and anti-Stokes radiation at low and high temperatures T.

6.4 For a linear triatomic molecule (e.g., CO_2) in a symmetric normal vibration, the two outer atoms oscillate toward one another; the atom in the middle remains at rest. Give a classical argument (dipole moment!) why no Raman scattering occurs here.

Section 6.3

6.5 Derive Eq. (6.40):
   (a) Using a stationary grating by considering the Bragg diffraction of light.
   (b) Using a moving grating by considering the Doppler effect.

6.6 For Brillouin scattering, the frequency shift $\omega_p - \omega_s$ is dependent on ω_p and the angle γ between the two optical wave vectors, but not for Raman scattering in molecules. What is the reason? *Hint*: Compare with Fig. 6.2.

6.7 Derive the solution, Eq. (6.45), for Brillouin scattering.

BIBLIOGRAPHY

G. P. Agrawal. *Nonlinear Fiber Optics*. Wiley, New York, 1984.

R. W. Boyd. *Nonlinear Optics*. Academic Press, San Diego, CA, 1992.

G. K. Grau. *Quantenelektronik*. Vieweg, Braunschweig, 1978.

M. Schubert and B. Wilhelmi. *Nonlinear Optics and Quantum Electronics*. Wiley, New York, 1986.

Y. R. Shen. *The Principles of Nonlinear Optics*. Wiley, New York, 1984.

A. Yariv. *Quantum Electronics*. Wiley, New York, 1975.

CHAPTER SEVEN

Optical Kerr Effect

The optical Kerr effect ($=$ quadratic electro-optic effect) is quite analogous to the Pockels effect (see Section 3.1): We use a field in order to induce optical anisotropy (birefringence) in an isotropic material or to change the anisotropy in an already birefringent crystal. Whereas in the linear electro-optic effect, the refractive index contains an additional term proportional to the electric field, for the quadratic electro-optic effect, this additional term is proportional to the square of the field. This effect, as four-wave mixing (FWM), can occur in all materials in principle, whereas the linear electro-optic effect, as a second-order effect, only occurs in media without an inversion center. Although such a FWM, in which a new field is generated from three fields, usually takes place effectively only if phase-matched (this easily can be fulfilled in general; see Chapter 8), the Kerr effect is independent of phase matching.

There are several physical mechanisms that produce an optically induced birefringence: The applied optical field can change the density of the material and/or the electronic charge distribution in the medium, leading to a change of the susceptibility and the index of refraction. $\chi^{(3)}$ can be resonantly amplified, if the change of frequency $|\omega' - \omega|$ lies close to some Raman transition (this is the basis of RIKES; see Section 8.3). If a linearly polarized beam propagates in a liquid, which consists of anisotropic molecules, the molecules will align due to the interaction of the induced dipoles with the field. The molecules will also be spatially redistributed to minimize the free energy of the system. This molecular reorientation and redistribution also lead to a change in the refractive index of the medium.

As applications, we discuss optical switches, bistability, and self-focusing.

7.1 GENERAL THEORY OF THE KERR EFFECT

For the special FWM process $f' = f' + f - f$ ($f = 0$), we find as polarization [see Eq. (1.61)]

$$\widehat{P}_i(f') = 3\varepsilon_0 \sum \chi_{ijkl}(-f', f', 0, 0)\widehat{E}_j(f')\widehat{E}_k(0)\widehat{E}_l(0) \qquad (7.1)$$

(a static field applied to a condenser).

In an isotropic medium, the $3^4 = 81$ components of the susceptibility tensor χ_{ijkl} are reduced to only three independent ones (see Appendix C): χ_{iijj}, χ_{ijij}, $\chi_{ijji}(i \neq j)$. We choose as representatives χ_{2211}, χ_{1212}, χ_{1221}. In addition, the subsidiary condition

$$\chi_{1111} = \chi_{2222} = \chi_{3333} = \chi_{2211} + \chi_{1212} + \chi_{1221}$$

holds.

Thus, we write [with $c^{(3)} = 3$]

$$\widehat{P}_i(f') = c^{(3)}\varepsilon_0 \sum_j [\chi_{iijj}(-f', f', 0, 0)\widehat{E}_i(f')\widehat{E}_j(0)\widehat{E}_j(0)$$

$$+ \chi_{ijij}(-f', f', 0, 0)\widehat{E}_j(f')\widehat{E}_i(0)\widehat{E}_j(0)$$

$$+ \chi_{ijji}(-f', f', 0, 0)\widehat{E}_j(f')\widehat{E}_j(0)\widehat{E}_i(0)] \qquad (7.2)$$

if for the polarization directions of the electrical fields $i \neq j$ holds. On the other hand, for $i = j$,

$$\widehat{P}_i(f') = c^{(3)} \cdot \varepsilon_0 \cdot \chi_{iiii}\widehat{E}_i(f')\widehat{E}_i(0)\widehat{E}_i(0) \qquad (7.3)$$

We assume that the pump field $\mathbf{E}(0)$ is polarized in the x-direction: $\widehat{\mathbf{E}}(0) = \widehat{E}(0)\mathbf{e}_x$. We are interested then in the polarizations $\widehat{P}_x(f') = \widehat{P}_1(f')$ and $\widehat{P}_y(f') = \widehat{P}_2(f')$, respectively:

$$\widehat{P}_x(f') = c^{(3)}\varepsilon_0\chi_{1111}\widehat{E}_x(f')\widehat{E}(0)^2$$

$$= c^{(3)}\varepsilon_0(\chi_{2211} + \chi_{1212} + \chi_{1221})\widehat{E}_x(f')\widehat{E}(0)^2 = \varepsilon_0\Delta\chi_{xx}\widehat{E}_x(f') \qquad (7.4)$$

$$\widehat{P}_y(f') = c^{(3)}\varepsilon_0\chi_{2211}\widehat{E}_y(f')\widehat{E}(0)^2 = \varepsilon_0\Delta\chi_{yy}\widehat{E}_y(f')$$

With this, the pump field-induced anisotropy in the susceptibility becomes

$$\delta\chi(f') = \Delta\chi_{xx} - \Delta\chi_{yy} = c^{(3)}(\chi_{1212} + \chi_{1221})\widehat{E}(0)^2 = \delta\varepsilon_r \qquad (7.5)$$

TABLE 7.1   Kerr constants K (at 20°C and for $\lambda = 0.589 \, \mu m$)

Benzene	0.67×10^{-14} m/V^2
Carbon disulfide	3.59×10^{-14} m/V^2
Chloroform	-8.85×10^{-14} m/V^2
Water	5.23×10^{-14} m/V^2
Chlorobenzene	11×10^{-14} m/V^2
Nitrotoluene	137×10^{-14} m/V^2
Nitrobenzene	245×10^{-14} m/V^2

However, the total dielectric constant at the frequency f' is

$$\varepsilon_r^{tot} = \varepsilon_r + \delta\varepsilon_r = n^2 + \delta\varepsilon_r = n_{tot}^2 = (n + \delta n)^2 \approx n^2 + 2n \cdot \delta n \qquad (7.6)$$

or

$$\delta n(f') = n_{\parallel} - n_{\perp} = \frac{\delta\chi(f')}{2n(f')} = \frac{c^{(3)}}{2n(f')}(\chi_{1212} + \chi_{1221})\widehat{E}(0)^2 \qquad (7.7)$$

The total index of refraction therefore has the form [see Eq. (A.6)]

$$n_{tot}(f') = n + \delta n, \qquad \delta n(f') = n_2 \cdot \widehat{E}(0)^2 = n_2' \cdot I(0) = n_2'' \cdot \mathcal{P}(0) \qquad (7.8)$$

and the additional term δn is proportional to the intensity $I(0)$, or the power $\mathcal{P}$ of the dc field. Note that the Kerr parameters n_2, n_2', n_2'' have different dimensions. Then the x- and y-components of the field $\widehat{E}(f')$ experience while propagating the distance $z = L$ a relative phase shift of

$$\delta\phi(f') = \frac{\omega'}{c} \cdot \delta n \cdot L - \frac{2\pi}{\lambda'} \cdot \delta n \; L \qquad (7.9)$$

This phase difference changes the state of polarization of the wave. Thus, an isotropic material becomes birefringent. Symmetry considerations show that the material will become optically uniaxial, and the optical axis points in the direction of the applied pump wave $\widehat{E}(0)$. Therefore, $n_{\parallel} = n_e$ (in the x-direction) yields the extraordinary beam $(\widehat{P}_x)$, $n_{\perp} = n_0$ the ordinary beam $(\widehat{P}_y)$.

Typical numerical values for n_2' are 10^{-20} to 10^{-18} m^2/W in glasses, 10^{-18} to 10^{-11} m^2/W in doped glasses, and 10^{-14} to 10^{-6} m^2/W in semiconductors.

We define a Kerr constant K via the relation [see Eq. (7.8), averaged over an optical period]

$$\delta n = K \cdot \lambda' \cdot |\widehat{E}|^2 \qquad (7.10)$$

In Table 7.1, we list Kerr constants for some liquids.

The optical Kerr effect occurs also in cases where the pump wave and $\widehat{E}(f')$ coincide ($f' = f \neq 0$), or where the pump wave has a very low frequency ($f \neq 0$).

7.2   APPLICATIONS

7.2.1   Optical Switches

An optical signal cannot pass a device containing a polarizer P_1 and an analyzer P_2 in crossed positions; see Fig. 7.1. However, if between P_1 and P_2 there is a Kerr cell, an intensive picosecond pump beam can rotate the polarization, and the light can pass the analyzer (partially). Here, nonlinear media with picosecond operation time are required. The Kerr effect is practically inertial less even for very rapidly changing fields and excellently suited as a highly resolving light shutter. The lower limit of the response time is given by the relaxation time τ of the molecules. (Nitrobenzene: $\tau = 4 \times 10^{-11}$ s, $CS_2 : \tau = 2 - 5 \times 10^{-12}$ s.)

Example 7.1   Let us consider transmitted intensity behind a Kerr optical shutter. We are given two polarization filters P_1 and P_2 with their transmission axes at right angles to each other and at $45°$ to the direction of the applied electric field ($= x$-direction). Calculate the dependence of the transmitted intensity of the optical field on the phase shift $\delta\phi$.

The incident plane vibrations behind the polarizer P_1 at $z = 0$, $E = a\cos\omega t$ (polarized along the $45°$ direction), with an initial intensity $I_0 \sim a^2$ can be broken up into two equal components in the x- and in y-direction:

$$E_x = \left(\frac{a}{\sqrt{2}}\right)\cos\omega t, \qquad E_y = \left(\frac{a}{\sqrt{2}}\right)\cos\omega t$$

At $z = L$, behind the plates of the capacitor but before P_2, we have

$$E_{x,L} = \left(\frac{a}{\sqrt{2}}\right)\cos\omega t, \qquad E_{y,L} = \left(\frac{a}{\sqrt{2}}\right)\cos(\omega t - \delta\phi)$$

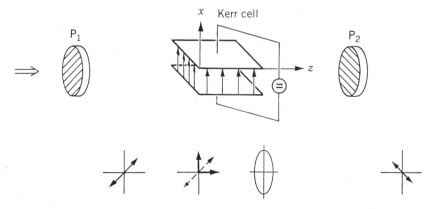

FIGURE 7.1   Kerr cell between a polarizer P_1 and an analyzer P_2 as electro-optic shutter.

up to a common constant phase. This represents elliptic polarization. Behind the analyzer P_2 (polarization along $-45°$), we have contributions w_1, w_2 from $E_{x,L}$ and $E_{y,L}$, respectively,

$$w_1 = 2^{-1/2} E_{x,L} = \left(\frac{a}{2}\right) \cos \omega t, \qquad w_2 = -2^{-1/2} E_{y,L} = -\left(\frac{a}{2}\right) \cos(\omega t - \delta\phi)$$

Therefore, the resulting vibration behind P_2 is given by

$$w = w_1 + w_2 = \left(\frac{a}{2}\right) [\cos \omega t \cdot (1 - \cos \delta\phi) - \sin \omega t \cdot \sin \delta\phi] = A \cos(\omega t + \psi)$$

where

$$A \cos \psi = \left(\frac{a}{2}\right)(1 - \cos \delta\phi), \qquad A \sin \psi = \left(\frac{a}{2}\right) \sin \delta\phi$$

For the transmitted intensity, we find

$$I_t \sim A^2 = a^2 \sin^2\left(\frac{\delta\phi}{2}\right), \qquad \frac{I_t}{I_0} = \sin^2\left(\frac{\delta\phi}{2}\right)$$

Alternative Derivation   Using the Jones matrices, we determine that, the Jones vector for the incident optical wave is $J_i = 1/\sqrt{2}\binom{1}{1}$, with polarization along the $45°$ direction. The Kerr cell works as a linear retarder and is represented by the 2×2 matrix:

$$T_K = \begin{bmatrix} 1 & 0 \\ 0 & \exp(-j\delta\phi) \end{bmatrix}$$

The Jones matrix for the analyzer P_2 has the form

$$T_A = \begin{bmatrix} \cos^2 \theta & \sin \theta \cos \theta \\ \sin \theta \cos \theta & \sin^2 \theta \end{bmatrix} = \frac{1}{2}\begin{bmatrix} 1 & -1 \\ -1 & 1 \end{bmatrix}$$

for polarization along $\theta = 135°$. Then, the Jones vector at the output is given by

$$J_o = T_A \cdot T_K \cdot J_i = \frac{1}{2\sqrt{2}}\begin{bmatrix} 1 - \exp(-j\delta\phi) \\ -1 + \exp(-j\delta\phi) \end{bmatrix} = \begin{bmatrix} a_1 \\ a_2 \end{bmatrix}$$

For the transmitted intensity, we find

$$I_t \sim |a_1|^2 + |a_2|^2 = \frac{|1 - \exp(-j\delta\phi)|^2}{4} = \sin^2\frac{\delta\phi}{2}$$

while the initial intensity is (from J_i) $I_0 \sim 1$. Thus, $I_t/I_0 = \sin^2(\delta\phi/2)$ again. ∎

7.2.2   Bistability

This means that for an input, there are two possibilities for the output signal—exactly as with magnetic hysteresis. Thus, for the correct processing of the input signal, the output signal must also be known, that is, all bistable systems require feedback (e.g. in a resonator), furthermore optical nonlinearity.

A commonly used realization in optics uses a Fabry–Perot resonator of length d, which contains a medium with an intensity-dependent refractive index of the form $n_{tot} = n + n_2'I$. In optics textbooks, it is shown that the intensity I_2 of the transmitted beam from the Fabry–Perot resonator is given by the Airy formula:

$$I_2 = \frac{T_0 I_1}{1 + F \cdot \sin^2(\delta/2)} = T \cdot I_1 \tag{7.11}$$

I_1 is the incoming intensity, T is the total transmission with $T_0 = T(\delta = 0)$, F is a constant connected with finesse, whereas the phase shift for a whole round-trip is

$$\delta = k_0 n_{tot} 2d = k_0(n + n_2'I_i)2d \tag{7.12}$$

and I_i is the intensity in the resonator. The output intensity I_2 can be expressed also with I_i:

$$I_2 = B \cdot I_i$$

where B depends on the mirror transmission; thus,

$$T = \frac{I_2}{I_1} = \frac{BI_i}{I_1} \tag{7.13}$$

The graphical solution of the two equations for $T = T(I_i)$, Eq. (7.11) plus Eq. (7.12) and Eq. (7.13), yields for small values of I_1 only one point of intersection in the curve $T(I_i)$, for larger I_1 values three, or still more points of intersection (see Fig. 7.2). From this result, we get $T(I_1)$ and $I_2(I_1)$. Bistability is very important for optical data processing in purely optical logic and computer systems. The switching effect is similar to phase transition. The nonlinear system has positive feedback and can lead to bifurcation and chaos in the output.

7.3   SELF-FOCUSING

Self-focusing is an effect that originates from the nonlinearity of the index of

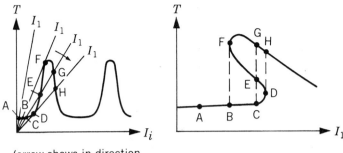

(arrow shows in direction
of increasing I_1 – values)

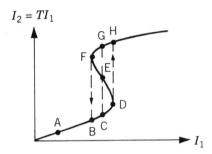

FIGURE 7.2   Bistability.

refraction:

$$n_{\text{tot}} = n + \delta n(I) \tag{7.14}$$

Laser light, which propagates in the z-direction and whose transversal intensity distribution has the form of a Gauss distribution, experiences the highest refractive index n_{tot} on the axis $x = y = 0$; for increasing distances $\rho = \sqrt{x^2 + y^2}$ from the axis, n_{tot} decreases. Optical path lengths are then maximum on the axis. Toward larger distances from the axis, the optical path lengths become shorter; the medium acts like a focusing lens. The laser light experiences a focusing effect that counteracts the expanding diffraction effect. Dependent on the ratio of the light power $\mathcal{P}_0$ to some critical power $\mathcal{P}_{cr}$, either diffraction or focusing predominates. The limiting case (neither diffraction nor focusing) is called *self-trapping*. Self-focusing is responsible for optical damages in high-power laser systems.

The effect of self-focusing is caused already by a single field

$$\hat{\mathbf{E}}(\mathbf{x}, f) = \mathbf{e}(f) \cdot \overline{E}(\mathbf{x}, f) \cdot \exp[-jk(f)z], \qquad k = k_0 n = \left(\frac{\omega}{c}\right)\sqrt{\varepsilon_r} \tag{7.15}$$

with frequency f [see Eq. (2.19), but now with a dependence also from the transversal coordinates x, y; the polarization unit vector $\mathbf{e}$ is directed along the x-direction]. The process $f = f - f + f$ is described by Eq. (1.60):

$$\sum_i \widehat{P}_i^{(3)}(f)e_i(f) = \varepsilon_0 \frac{3}{4} \sum \chi_{ijkl}(-f, f, -f, f)\widehat{E}_j(f)\widehat{E}_k^*(f)\widehat{E}_l(f) \cdot e_i(f)$$

$$= \varepsilon_0 \frac{3}{4} \Big[\sum \chi_{ijkl}(-f, f, -f, f)e_j(f)e_k^*(f)e_l(f) \cdot e_i(f)$$

$$\cdot \overline{E}(\mathbf{x}, f)\overline{E}^*(\mathbf{x}, f) \Big] e^{-jk(f)z} \overline{E}(\mathbf{x}, f)$$

$$= \varepsilon_0 \cdot \chi_{\text{eff}}[f, |\overline{E}(\mathbf{x}, f)|^2] \cdot \overline{E}(\mathbf{x}, f)e^{-jk(f)z} \qquad (7.16)$$

with an effective susceptibility

$$\chi_{\text{eff}}[f, |\overline{E}(x, f)|^2] := \frac{3}{4} \Big[\sum \chi_{ijkl}(-f, f, -f, f)e_i e_j e_k^* e_l \Big] |\overline{E}(\mathbf{x}, f)|^2$$

$$=: \frac{\kappa}{k_0^2} |\overline{E}|^2 =: \varepsilon_r^{nl} \qquad (7.17)$$

The formalism of Section 2.1.3 has to be extended to fields that possess a weak dependence on the transverse coordinates x, y. Then we get with $\text{div } \mathbf{E} = \partial_x E_x \approx 0$

$$\nabla \times (\nabla \times \mathbf{E}) \approx -\nabla^2 \mathbf{E} = -(\partial_x^2 + \partial_y^2 + \partial_z^2)\mathbf{E} = -(\nabla_t^2 + \partial_z^2)\mathbf{E} \qquad (7.18)$$

Thus, there is now an additional term $-\nabla_t^2 \mathbf{E}$ compared with the former treatment, and we obtain with Eqs. (7.16) and (7.17) instead of Eq. (2.21)

$$2jk\partial_z \overline{E}(\mathbf{x}, f) = \nabla_t^2 \overline{E}(\mathbf{x}, f) + \omega^2 \mu_0 \varepsilon_0 \cdot \frac{1}{\varepsilon_0} \sum \widehat{P}_i(\mathbf{x}, f)e_i(f)e^{jk(f)z}$$

$$= \nabla_t^2 \overline{E}(\mathbf{x}, f) + k_0^2 \chi_{\text{eff}} \overline{E}(\mathbf{x}, f) = [\nabla_t^2 + k_0^2 \varepsilon_r^{nl}(|\overline{E}|^2)] \cdot \overline{E}(\mathbf{x}, f) \quad (7.19)$$

This parabolic differential equation corresponds to a time-dependent *nonlinear Schrödinger equation* if the coordinate z is replaced by time t. Setting

$$q = \frac{z}{(2k)}$$

we get with κ from Eq. (7.17) a normalized form of the PDE

$$j \cdot \frac{\partial \overline{E}}{\partial q} = \nabla_t^2 \overline{E} + \kappa \cdot |\overline{E}|^2 \overline{E} \qquad (7.20)$$

In the case in which $\overline{E}(\mathbf{x}, f_1)$ depends on x and y, no closed solution is known.

If, however, there is no dependence on y, that is, $\overline{E}(\mathbf{x}, f) = \overline{E}(x, z, f)$, Eq. (7.20) can be solved. By inspection, one can see that for $\kappa > 0$ (see Appendix E)

$$\overline{E}(x, z, f) = \sqrt{\frac{2}{\kappa}} \cdot \eta \cdot \frac{\exp[j(\xi^2 - \eta^2)q + j\xi x]}{\cosh[\eta(x - x_0) + 2\eta\xi q]} \tag{7.21}$$

is a solution. Here, ξ, η, and x_0 are three undetermined constants. A phase term $\exp(-j\phi)$ is still possible, but will be ignored in the following: $\phi = 0$.

To begin with, we set $\xi = 0$ and obtain the special solution

$$\overline{E}(x, z, f) \cdot e^{-jkz} = \sqrt{\frac{2}{\kappa}} \cdot \eta \cdot \frac{\exp(-j\eta^2 q)}{\cosh[\eta(x - x_0)]} \cdot e^{-jkz} \tag{7.22}$$

This beam has an x-dependent profile, the beam radius (defined by the $1/e$ value) is approximately $1/\eta$, since $\cosh \eta u = (e^{\eta u} + e^{-\eta u})/2 \approx e^{\eta u}/2$, and the propagation constant is $k + \eta^2/(2k)$.

If, however, $\xi \neq 0$, the beam center lying at $x = x_0$ for $z = 0$ is shifted to $x_0 - \xi z/k$, that is, the beam is tilted against the z-axis by an angle

$$\theta \approx \tan \theta = \frac{\xi}{k}$$

Combined with this is a tilt of the phase fronts, and the total phase is now given by

$$\exp\left\{-jk\left(1 + \frac{\eta^2 - \xi^2}{2k^2}\right)z + j\xi x\right\} = \exp(-j\mathbf{k}^{nl} \cdot \mathbf{x})$$

$$\mathbf{k}^{nl} = \left\{-\xi, 0, k\left(1 + \frac{\eta^2 - \xi^2}{2k^2}\right)\right\}$$

that is, the total propagation vector has a tilt angle ϑ against the z-axis defined by $\mathbf{k}^{nl} \cdot \mathbf{e}_z = |\mathbf{k}^{nl}| \cdot \cos \vartheta$. Thus, with $\xi^2 \approx \eta^2 \ll k^2$,

$$\cos \vartheta = \frac{\mathbf{k}^{nl} \cdot \mathbf{e}_z}{|\mathbf{k}^{nl}|} = \frac{2k^2 + \eta^2 - \xi^2}{\sqrt{[4k^4 + 4k^2\eta^2 + (\eta^2 - \xi^2)^2]}} \approx \frac{1 + (\eta^2 - \xi^2)/(2k^2)}{1 + \eta^2/(2k^2)} \approx 1 - \frac{\vartheta^2}{2}$$

or

$$\vartheta \approx \frac{\xi}{k} = \theta \tag{7.23}$$

and the tilt angles of the beam center and the propagation vector coincide approximately.

In Eq. (7.20), we now use the ansatz (with dependence on x and y)

$$\overline{E}(\mathbf{x}) = A_0(\mathbf{x}) \cdot \exp[-jkS(\mathbf{x})] \tag{7.24}$$

with real quantities A_0 and S ($=$ eikonal) and obtain after separation into real and imaginary parts the two equations

$$2 \cdot \frac{\partial S}{\partial z} + (\nabla_t S)^2 = \frac{(\nabla_t^2 A_0)}{(k^2 A_0)} + \frac{\varepsilon_r^{nl}}{\varepsilon_r} \tag{7.25}$$

$$\frac{\partial A_0^2}{\partial z} + \nabla_t(A_0^2 \nabla_t S) = 0 \tag{7.26}$$

Equation (7.26) corresponds to an equation of continuity and represents the conservation of energy. Equation (7.25) shows how the wave front, represented by the eikonal S, will be distorted by two influences: a diffraction effect (first term on the right-hand side) and a nonlinear effect (second term).

We look for solutions where both effects just cancel [i.e., the right-hand side of Eq. (7.25) vanishes] and that provide a constant profile: $\nabla_t S = 0$. Then we get $\partial S/\partial z = 0$ and $\partial A_0/\partial z = 0$. This is the case of self-trapping. The wave propagates in a medium with, for example, plane wave fronts and with a transverse profile constant with respect to z: *Spatial solitons* are created (see Fig. 7.3a). In fact, this solution is very unstable; a small decrease of laser power due to absorption or scattering can again destroy the balance between diffraction and nonlinearity. The analytic solutions of Eqs. (7.25) and (7.26) are known only in trivial cases.

With the help of a simple consideration, we acquire an approximate expression for the critical power $\mathcal{P}_{cr}$ for self-trapping. For simplicity, we consider a transverse intensity distribution having a constant value within a circle with diameter d (in the xy-plane), but vanishing at the outside. The medium acts then as a cylindrical waveguide with an increased refractive index $n_{tot} = n + \delta n$, embedded in a medium with a refractive index n. We assume that the laser light consists of a whole bundle of rays guided in the waveguide by total internal reflection. Total reflection due to this nonlinear guiding occurs if the angle θ against the boundary plane (or against the z-axis) is smaller than the limiting angle θ_0, for which

$$\cos \theta_0 = \frac{n}{n + \delta n} \approx 1 - \frac{\delta n}{n} \approx 1 - \frac{\theta_0^2}{2}$$

holds, or

$$\theta_0 \approx \sqrt{2 \cdot \frac{\delta n}{n}} \tag{7.27}$$

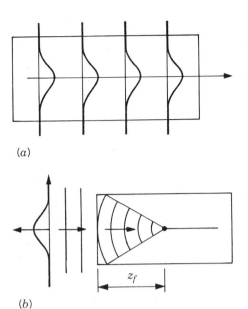

(a)

(b)

FIGURE 7.3 (*a*) Self-trapping with spatial solitons, (*b*) self-focusing. (Reproduced with permission from B.E.A. Saleh and M.C. Teich, *Fundamentals of Photonics*, John Wiley & Sons, Inc., New York, 1991.)

Vice versa, light experiences diffraction at a circular aperture of radius $d/2$, and the angle (against the z-axis) extended by the central disk in Fraunhofer diffraction is given by

$$\tan\theta_d = \frac{1.22\lambda}{nd}, \qquad \theta_d \approx \frac{1.22\lambda}{nd} \qquad (7.28)$$

We expect self-trapping if rays making this diffraction angle with respect to the z-axis are still totally reflected due to nonlinearity, or if $\theta_0 = \theta_d$, or if the critical intensity I_{cr} satisfies [see Eq. (7.8)]

$$n_2' I_{cr} = \delta n = \frac{1}{2n} \cdot \left(\frac{1.22\lambda}{d}\right)^2 \qquad (7.29)$$

From this critical intensity results critical power in the cross section $(\pi/4) \cdot d^2$:

$$\mathcal{P}_{cr} = \frac{\pi}{4} \cdot d^2 I_{cr} = \frac{\pi}{8nn_2'}(1.22\lambda)^2 \qquad (7.30)$$

being independent of the beam cross section. $\mathcal{P}_{cr}$ lies at some 10 kW in liquids with strong nonlinearity and at some few MW for weak nonlinearity.

For a laser light power $\mathcal{P}_0 > \mathcal{P}_{cr}$, nonlinearity prevails and the beam will be focused on a point at a certain distance z_f; see Fig. 7.3b. In order to estimate this distance z_f, we use again a transverse intensity distribution, but now with a Gaussian outward slope [see Eq. (4.10)] with a radius $w_0 = d/2$ (given by a decrease of the amplitude to $1/e$). At $z = 0$, the nonlinear medium starts. The geometrical path length is z_f on the axis, at the periphery about $(z_f^2 + w_0^2)^{1/2}$, with an averaged index of refraction of $n + \delta n/2$ there. The optical paths must now coincide along both paths; thus, with $z_f \gg w_0$,

$$z_f(n + \delta n) = \sqrt{z_f^2 + w_0^2} \cdot \left(n + \frac{\delta n}{2}\right) \approx z_f\left(1 + \frac{w_0^2}{2z_f^2} + \dots\right)\left(n + \frac{\delta n}{2}\right)$$

So, we find approximately

$$z_f \approx \sqrt{\frac{n}{\delta n}} \cdot w_0 = \sqrt{2} \cdot \frac{w_0}{\theta_0} = \sqrt{\frac{n}{In_2'}} \cdot w_0 = \sqrt{\frac{n\pi}{P_0 n_2'}} \cdot w_0^2 \qquad (7.31)$$

with $P_0 = \pi w_0^2 \cdot I$. The elimination of n_2' using Eq. (7.30) yields finally

$$z_f = \frac{\sqrt{8}nw_0^2}{1.22\lambda} \cdot \sqrt{\frac{P_{cr}}{P_0}} \qquad (7.32)$$

Let us write the z-dependent radius of the Gaussian beam in the form $w_0 \cdot F(z)$. With a more exact theory, one finds then

$$F^2(z) = 1 - \left(\frac{P_0}{P_{cr}} - 1\right) \cdot \left(\frac{z}{w_0^2 k_0}\right)^2 \qquad (7.33)$$

If the laser light power $P_0 \ll P_{cr}$, one obtains the usual diffraction divergence of a Gaussian beam. For $P_0 = P_{cr}$, we get $F(z) \equiv 1$: This is the case of self-trapping. Finally, for $P_0 > P_{cr}$, the beam has shrunk to a point at a distance of

$$z = z_f = \frac{k_0 w_0^2}{\sqrt{P_0/P_{cr} - 1}} \approx k_0 w_0^2 \sqrt{\frac{P_{cr}}{P_0}} \qquad (7.34)$$

Thus, $w_0 F(z_f) = 0$; the beam is focused. This agrees approximately with the simple formula of Eq. (7.32).

PROBLEMS

Section 7.1

7.1 What has to be changed in Section 7.1 if instead of a static field a low-frequency field is applied?

7.2 For the Kerr effect:
   (a) Calculate δn for $K = 5 \times 10^{-12}$ m/V^2, $\lambda = 589$ nm, $|E| = 3 \times 10^5$ V/m.
   (b) What is the phase shift $\delta\phi$, if the length of the electrodes is 4 cm?

7.3 Determine the potential difference required between the plates of a Kerr cell (filled with nitrotoluene) that are 5 cm long and 2 cm apart in order to restore full brightness through the crossed polarizers of the Kerr optical shutter.

7.4 The plates of a Kerr optical shutter are 5 cm long and 0.7 cm apart and kept at a potential difference of 10,000 V. The cell is filled first with carbon disulfide and then nitrobenzene. Compute the fractions of the incident intensities passing through the polarization filters and the cell in the two cases.

7.5 Determine the field strength required to produce circular polarization ($\delta\phi = \pi/2$) by a Kerr cell filled with CS$_2$ if the path length is 10 cm.

Section 7.3

7.6 In an experiment on the self-focusing of a ruby laser beam ($\lambda = 0.6943$ μm) in carbon disulfide, a critical power $P_{cr} \approx 25$ kW was found. Compute the critical power using Eq. (7.30) with $n = 1.628$ and $n_2' = 4.63 \times 10^{-18}$ m^2/W. Even this rough calculation gives the correct order of magnitude.

BIBLIOGRAPHY

G. P. Agrawal. *Nonlinear Fiber Optics*. Wiley, New York, 1984.

F. T. Arecchi and E. O. Schulz-Dubois (eds.). *Laser Handbook*. Vols. 1 and 2. North-Holland. Amsterdam, 1972.

R. W. Boyd. *Nonlinear Optics*. Academic Press, San Diego, CA, 1992.

H. H. A. Haus. *Waves and Fields in Optoelectronics*. Prentice-Hall, Englewood Cliffs, NJ, 1984.

B. E. A. Saleh and M. C. Teich. *Fundamentals of Photonics*. Wiley, New York, 1991.

M. Schubert and B. Wilhelmi. *Nonlinear Optics and Quantum Electronics*. Wiley, New York, 1986.

Y. R. Shen. *The Principles of Nonlinear Optics*. Wiley, New York, 1984.

A. Yariv. *Quantum Electronics*. Wiley, New York, 1975.

F. Zernike and J. E. Midwinter. *Applied Nonlinear Optics*. Wiley, New York, 1973.

CHAPTER EIGHT

Four-Wave Mixing

In this chapter, we treat some general features and also some particular processes of four-wave mixing (FWM), having already discussed as special cases of FWM in Chapter 6 the induced Raman effect and in Chapter 7 the optical Kerr effect. The special cases we shall discuss are phase conjugation, nonlinear spectroscopy, third harmonic generation, and two-photon absorption.

In FWM, four interacting electromagnetic fields are participating: Three fields produce a fourth field. We have a third-order process $P^{(3)}$ with a nonlinear susceptibility $\chi_{ijkl}^{(3)}$. FWM is the lowest-order nonlinearity for media with inversion symmetry. For media without inversion symmetry, the process is, in general, much weaker than three-wave mixing since $|\chi^{(3)} \cdot (V/m)^2| \ll |\chi^{(2)} \cdot V/m|$; nevertheless, the process can easily be observed at high laser intensities or for long interaction lengths.

8.1 GENERAL THEORY OF FOUR-WAVE MIXING

We use the following simplifying assumptions: We assume a cubic or isotropic medium; a constant pump (no pump depletion); the nonlinear medium is in the half-space $z \geq 0$; the propagation of the signal wave at f_s takes place in the z-direction. In Fig. 8.1, the four cases to be discussed are plotted.

1. In the case of three pump fields $\mathbf{E}_1(\mathbf{x}, t)$, $\mathbf{E}_2(\mathbf{x}, t)$, $\mathbf{E}_3(\mathbf{x}, t)$, we set for the monochromatic fields [complex analytic signals, see Eq. (2.19)]

$$\widetilde{\mathbf{E}}_r(\mathbf{x}, t) = \mathbf{e}_r \overline{E}(z, f_r) \cdot \exp j(\omega_r t - \mathbf{k}_r \cdot \mathbf{x}), \qquad r = 1, 2, 3 \qquad (8.1)$$

whereas the output field (signal field, subscript s), produced by the polarization $P^{(3)}(f_s)$, is

$$\widetilde{\mathbf{E}}_s(\mathbf{x}, t) = \mathbf{e}_s \overline{E}(z, f_s) \cdot \exp j(\omega_s t - \mathbf{k}_s \cdot \mathbf{x}), \qquad f_s = f_1 + f_2 + f_3 \qquad (8.2)$$

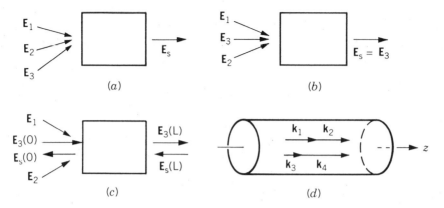

FIGURE 8.1   Four waving-mixing: (*a*) three pump waves E_1, E_2, E_2; (*b*) two pump waves E_1, E_2 and $E_3 = E_s$; (*c*) parametrically produced signal and idler wave (E_3); and (*d*) partially degenerated four wave-mixing in glass fibers. (Reproduced with permission from Y. R. Shen, *The Principles of Nonlinear Optics*, John Wiley & Sons, Inc., New York, 1984.)

In the approximation of the slowly varying amplitude, Eq. (2.21),

$$\frac{d}{dz}\overline{E}_s(z,f_s) = -j\frac{\omega_s^2 \mu_0}{2k_s}\sum_i \widehat{P}_i(z,f_s)e_{si}(f_s)\cdot \exp(jk_s z) \tag{8.3}$$

We have to use third-order polarization; similarly to Eq. (2.22), we write with $\widehat{\mathbf{P}} = \widehat{P}(z,f_s)\cdot \mathbf{e}_s$

$$\widehat{P}_i(z,f_s) = e_{si}\widehat{P}(z,f_s)$$

$$= c^{(3)}\varepsilon_0 \sum \chi_{ijkl}(-f_s,f_1,f_2,f_3)e_{1j}e_{2k}e_{3l}$$

$$\times \overline{E}(f_1)\overline{E}(f_2)\overline{E}(f_3)\cdot \exp[-j(\mathbf{k}_1 + \mathbf{k}_2 + \mathbf{k}_3)\cdot \mathbf{x}] \tag{8.4}$$

With the effective susceptibility

$$\chi_{\text{eff}}^{(3)} = c^{(3)}\sum \chi_{ijkl}\cdot e_{si}e_{1j}e_{2k}e_{3l} \tag{8.5}$$

we obtain

$$\widehat{P}(z,f_s) = \varepsilon_0 \chi_{\text{eff}}^{(3)}\overline{E}_1(z,f_1)\overline{E}_2(z,f_2)\overline{E}_3(z,f_3)\cdot \exp[-j(\mathbf{k}_1 + \mathbf{k}_2 + \mathbf{k}_3)\cdot \mathbf{e}_z z] \tag{8.6}$$

Then, we get from the differential equation (8.3)

$$\frac{d}{dz}\overline{E}(z,f_s) = -j\frac{\omega_s^2 \mu_0 \varepsilon_0}{2k_s}\chi_{\text{eff}}^{(3)}\overline{E}(z,f_1)\overline{E}(z,f_2)\overline{E}(z,f_3)\cdot \exp[-j(\Delta \mathbf{k}\cdot \mathbf{e}_z)z] \tag{8.7}$$

and the wave vector mismatch (phase mismatch) is

$$\Delta k = k_1 + k_2 + k_3 - k_s \tag{8.8}$$

Integration with the initial condition $\overline{E}(z=0, f_s) = 0$ and with $\overline{E}(0, f_1)$, $\overline{E}(0, f_2)$, $\overline{E}(0, f_3)$ as constant pump waves in case 1 results in

$$\overline{E}(z, fs) = \frac{\omega_s^2 \mu_0 \varepsilon_0}{2 k_s (\Delta \mathbf{k} \cdot \mathbf{e}_z)} \chi_{\text{eff}}^{(3)} \overline{E}(0, f_1) \overline{E}(0, f_2) \overline{E}(0, f_3) \cdot \left[e^{-j(\Delta \mathbf{k} \cdot \mathbf{e}_z)z} - 1 \right] \tag{8.9}$$

We have a modulation of power:

$$|\overline{E}_s|^2 \sim z^2 \cdot \frac{\sin^2\left(\dfrac{\Delta \mathbf{k} \cdot \mathbf{e}_z}{2} z\right)}{\left(\dfrac{\Delta \mathbf{k} \cdot \mathbf{e}_z}{2} z\right)^2}$$

Therefore, phase matching $\Delta k = 0$ is decisive too. However, this can be achieved in FWM in a large number of ways by an appropriate alignment of the directions of propagation of the three pump waves. For $\Delta k = 0$, we find an increase in intensity $\sim z^2$.

2. We use now only two (nondepleted) pump fields $\overline{E}_1 \equiv \overline{E}(0, f_1)$, $\overline{E}_2 \equiv \overline{E}(0, f_2)$; the output field $\overline{E}_s$ is in the same mode as $\overline{E}_3 : \overline{E}(z, f_s) = \overline{E}_3(z, f_3)$, $f_s = f_3$, $\mathbf{k}_s = \mathbf{k}_3$, $f_1 = -f_2$, $\Delta \mathbf{k} \to \Delta \mathbf{k}' = \mathbf{k}_1 + \mathbf{k}_2$. Then, we get from Eq. (8.7) after integration

$$\overline{E}(z, f_s) = \overline{E}(0, f_s) \cdot \exp g_s(z) \tag{8.10}$$

with

$$g_s(z) = \frac{\omega_s^2 \mu_0 \varepsilon_0}{2 k_s (\Delta \mathbf{k}' \cdot \mathbf{e}_z)} \chi_{\text{eff}}^{(3)} \overline{E}(0, f_1) \overline{E}(0, f_2) \cdot [e^{-j(\Delta \mathbf{k}' \cdot \mathbf{e}_z)z} - 1] \tag{8.11}$$

The real part of g_s represents gain or loss. With,

$$\overline{E}(0, f_2) = \overline{E}(0, -f_1) = \overline{E}^*(0, f_1)$$

we get for small values of z

$$\text{Re}[g_s(z)] \approx \frac{\omega_s^2 \mu_0 \varepsilon_0}{2 k_s} \text{Im}\left[\chi_{\text{eff}}^{(3)} \right] \cdot |\overline{E}(0, f_1)|^2 \cdot z \tag{8.12}$$

The exponential growth is similar to that of the Stokes wave of the Raman effect.

3. In parametric backward wave amplification or oscillation, two strong fields $\overline{E}_1 = \overline{E}(0,f_1)$ and $\overline{E}_2 = \overline{E}(0,f_2)$ act as constant pump waves, while the two weak contradirectional waves $\overline{E}_s = \overline{E}(z,f_s)$ and $\overline{E}_3 = \overline{E}(z,f_3)$ (idler wave) become amplified. We have the same situation as for the usual parametric amplification (see Sections 5.4 and 5.5), except that now two pump waves occur instead of only one. Besides Eq. (8.7), one has to write down also the corresponding differential equation for $\overline{E}_3$ (*Note:* $f_3 = f_s - f_1 - f_2$):

$$\frac{d}{dz}\overline{E}(z,f_3) = -j\frac{\omega_3^2\mu_0\varepsilon_0}{2k_3}\chi_{\text{eff}}^{(3)}\overline{E}(z,f_s)\overline{E}^*(0,f_1)\overline{E}^*(0,f_2)\cdot\exp[j(\Delta\mathbf{k}\cdot\mathbf{e}_z)z] \quad (8.13)$$

for a lossless system. We solve the system of Eqs. (8.7) and (8.13) for $\Delta\mathbf{k} = 0$. The elimination of $\overline{E}_3$ and $\overline{E}_s$, respectively, yields

$$\left(\frac{d^2}{dz^2}+\kappa^2\right)\cdot\begin{bmatrix}\overline{E}_s\\\overline{E}_3\end{bmatrix} = 0, \qquad \kappa^2 = \frac{\omega_s^2\omega_3^2}{k_sk_3}\cdot\left|\frac{\chi_{\text{eff}}^{(3)}\overline{E}_1\overline{E}_2}{2c^2}\right|^2 \quad (8.14)$$

The initial conditions are given values for $\overline{E}_s(z = L)$ and $\overline{E}_3(z = 0)$. We look for $\overline{E}_s(z = 0)$ and $\overline{E}_3(z = L)$. Explicitly, we find easily with $A = \chi_{\text{eff}}^{(3)}\cdot\overline{E}_1\overline{E}_2/(2c^2)$

$$\overline{E}(z = 0,f_s) = \frac{\overline{E}(L,f_s)}{\cos\kappa L}+j\frac{A}{|A|}\frac{\omega_s}{\omega_3}\sqrt{\frac{k_3}{k_s}}\overline{E}(0,f_3)\cdot\tan\kappa L$$

$$\overline{E}(z = L,f_3) = -j\frac{|A|}{A}\frac{\omega_3}{\omega_s}\sqrt{\frac{k_s}{k_3}}\overline{E}(L,f_s)\cdot\tan\kappa L+\frac{\overline{E}(0,f_3)}{\cos\kappa L} \quad (8.15)$$

If $\kappa L \to \pi/2$, then $\overline{E}_s(z = 0)$ and $\overline{E}_3(z = L)$ diverge. This means that even for $\overline{E}_s(L) = 0$ and $\overline{E}_3(0) = 0$ (without any input field) fields $\overline{E}_s(0)$, $\overline{E}_3(L)$ can be generated: One has the generation of oscillations. This is then a parametric oscillator. Of course, power is taken from both pump waves $\overline{E}_1$, $\overline{E}_2$.

Whereas for this FWM phase matching is easily realizable, the conventional parametric backward wave oscillator with only one pump wave, where signal and idler waves are contradirectional, has not yet been observed experimentally. This is due to the difficulty in guaranteeing here phase matching.

4. For the process $f_4 = f_1 + f_2 + f_3$ (possibly with $f_1 = f_2 = f_3$ or $f_1 = f_2 \neq f_3$), conservation of momentum (phase matching) is, in general, difficult to satisfy in glass fibers where all momentum vectors are parallel to the fiber axis. In contrast, this is easy to achieve in the "partially degenerate" case:

$$f_4 = f_1 + f_2 - f_3 = 2f_p - f_3$$

$$\Delta k = k_4 + k_3 - k_2 - k_1 = \frac{2\pi(f_4n_4 + f_3n_3 - f_2n_2 - f_1n_1)}{c} \quad (8.16)$$

with $k_i = 2\pi f_i n_i/c$. We set here $f_1 = f_2 = f_p$ (pump frequency) and $k_1 = k_2 = k_p$. Assume that all waves propagate in the $+z$-direction and choose, without restricting generality, $f_4 > f_3$. (Previously, this partially degenerate FWM has been denoted sometimes as three-wave mixing since only three different frequencies occur).

A strong pump at $f_1 = f_p = (f_3 + f_4)/2$ generates the two symmetrical sidebands at f_3 and f_4 with a frequency shift of

$$\Omega_s = \omega_p - \omega_3 = \omega_4 - \omega_p \tag{8.17}$$

f_3 will be called the Stokes frequency in analogy to the Raman effect; f_4 is the anti-Stokes frequency. If only the pump wave is incident in the glass fiber, the frequencies f_3, f_4 will be spontaneously generated from noise fluctuations if the conservation of momentum is fulfilled. However, if initially there is already a weak signal at f_4, the signal is amplified and an idler wave at f_3 is generated (stimulated FWM in contrast to spontaneous FWM). We shall meet this situation again in Section 9.2.4 (modulation instability).

For a constant pump, one has to write down two differential equations, similar in type to Eqs. (8.7) and (8.13), for $f_3 = 2f_p - f_4$ and $f_4 = 2f_p - f_3$:

$$\frac{d\overline{E}(z,f_3)}{dz} = -j\frac{\omega_3^2 \mu_0 \varepsilon_0}{2k_3} \chi_{\text{eff}}^{(3)} \overline{E}^2(0,f_1) \overline{E}^*(z,f_4) \cdot e^{j\Delta k \cdot z}$$

$$\frac{d\overline{E}(z,f_4)}{dz} = -j\frac{\omega_4^2 \mu_0 \varepsilon_0}{2k_4} \chi_{\text{eff}}^{(3)} \overline{E}^2(0,f_1) \overline{E}^*(z,f_3) \cdot e^{j\Delta k \cdot z} \tag{8.18}$$

with

$$\chi_{\text{eff}}^{(3)} = c^{(3)} \sum \chi_{ijkl}(-f_4,f_1,f_1,-f_3)e_j(f_1)e_k(f_1)e_l(f_3)e_i(f_4)$$

$$= c^{(3)} \sum \chi_{ijkl}(-f_3,f_1,f_1,-f_4)e_j(f_1)e_k(f_1)e_l(f_4)e_i(f_3)$$

Elimination of, for example, $\overline{E}(z,f_4)$ from Eq. (8.18) yields for a constant pump a second-order ODE

$$\overline{E}''(z,f_3) - j\Delta k \cdot \overline{E}'(z,f_3) - K^2\overline{E}(z,f_3) = 0$$

with

$$K = \frac{\omega_3\omega_4\mu_0\varepsilon_0\left|\chi_{\text{eff}}^{(3)}\right|\left|\overline{E}(0,f_1)\right|^2}{2\sqrt{k_3k_4}} \tag{8.19}$$

As the solution of this differential equation, we get for $\overline{E}(z,f_3)$ [and similarly for

$\overline{E}(z, f_4)$]:

$$\overline{E}(z, f_3) = \exp\left(+j\Delta k \cdot \frac{z}{2} \right) \cdot (N_1 e^{gz} + N_2 e^{-gz})$$

$$\overline{E}^*(z, f_4) = \exp\left(-j\Delta k \cdot \frac{z}{2} \right) \cdot (N_3 e^{gz} + N_4 e^{-gz})$$

(8.20)

with a parametric gain

$$g = \sqrt{K^2 - \left(\frac{\Delta k}{2}\right)^2}$$

(8.21)

For glass fibers, this expression must still be modified and discussed. This will be done in Section 11.1.

If pump depletion is taken into account, one must replace $E(0, f_1)$ in Eqs (8.18) by $E(z, f_1)$. These equations must then be solved numerically in general.

8.2 DEGENERATE FOUR-WAVE MIXING; PHASE CONJUGATION

All waves now have the same frequency f_0: $f_0 = f_0 + f_0 - f_0$, that is, one of the three frequencies f_1, f_2, f_3 must be set equal to $-f_0$. We choose $f_3 = -f_0$. Hence, three different configurations are imaginable, from which, in general, only a single one is realized due to the phase-matching condition. One can treat the process with the theory outlined in Section 8.1. But it can also be described by the following simple physical picture: Two of the three input waves interfere and produce in the nonlinear medium a static grating remaining fixed in space or a propagating one. The third input wave is diffracted from this grating and so produces the output wave; see also Problem 6.5.

Degenerate FWM is an important method to generate a phase-conjugated wave (other methods are difference frequency generation, parametric amplification, stimulated Brillouin scattering, stimulated Raman scattering). *Phase conjugation* is defined as follows: An incident wave

$$E_1(\mathbf{x}, t) = \mathrm{Re}\left[\overline{E}(\mathbf{x}) \cdot \exp j(\omega t - kz) \right]$$

(8.22)

is transformed by the process of phase conjugation into the phase-conjugated wave:

$$E_2(\mathbf{x}, t) = \mathrm{Re}\left[\overline{E}^*(\mathbf{x}) \cdot \exp j(\omega t + kz) \right] = \mathrm{Re}\left[\overline{E}(\mathbf{x}) \cdot \exp j(-\omega t - kz) \right]$$

(8.23)

This wave propagates backward, and the phase in $\overline{E}(\mathbf{x}) = |\overline{E}| \cdot \exp(j \arg \overline{E})$ simultaneously changes sign. This is equivalent to time reversal ($t \rightarrow -t$). The

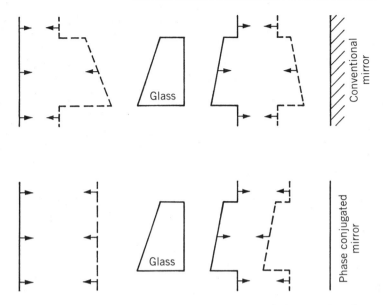

FIGURE 8.2   Phase fronts for mirror reflection and phase conjugation.

phase-conjugated wave is not identical to the reflected wave, where $\overline{E}(\mathbf{x})$ remains unchanged; see Fig. 8.2. Now, the law of reflection does not hold (the wave is reflected back onto itself no matter what the angle of incidence is), and a divergent beam runs backward as a convergent beam while from a conventional mirror a divergent beam is reflected divergently.

Let the waves $\overline{E}_s$ and $\overline{E}_3$ from the FWM now propagate contradirectionally:

$$\mathbf{k}_s = -\mathbf{k}_3, \qquad |\mathbf{k}_s| = |\mathbf{k}_3| =: k; \qquad f_s = f_0 = -f_3$$

$$\overline{E}_3(z,f_3) = \overline{E}_3(z,-f_0) = \overline{E}_3^*(z,f_0)$$

If we set still $\overline{E}_s(z = L, f_0) = 0$, we get from the solution of Eq. (8.15)

$$\overline{E}_s(z = 0, f_0) = -j\frac{A}{|A|} \tan \kappa L \cdot \overline{E}_3^*(z = 0, f_0) =: r \cdot \overline{E}_3^*(z = 0, f_0) \qquad (8.24)$$

But together with $\mathbf{k}_s = -\mathbf{k}_3$, this means phase conjugation. The nonlinear medium for the degenerate FWM serves simply as a phase-conjugating mirror. The phase-conjugated wave can be even more intense as the incident wave:

$$|\overline{E}_s|^2 > |\overline{E}_3|^2, \qquad \text{if} \quad \frac{\pi}{4} < \kappa L < \frac{3\pi}{4}$$

and oscillation starts, if $\kappa L \to \pi/2$. If both pump waves (with $|\overline{E}_1| = |\overline{E}_2|$) also propagate contradirectionally, $\mathbf{k}_1 = -\mathbf{k}_2$. Then from phase matching immediately follows the desired relation:

$$\mathbf{k}_s + \mathbf{k}_3 = \mathbf{k}_1 + \mathbf{k}_2 = 0$$

From Eq. (8.14), we find ($k = \omega_0 n_0 / c = k_0 n_0$)

$$\kappa = \frac{\omega_0^2}{k} |A|^2 = \frac{\omega_0^2}{2kc^2} |\chi_{\text{eff}}^{(3)}| \cdot |\overline{E}_1|^2 = \frac{k_0}{2n_0} |\chi_{\text{eff}}^{(3)}| \cdot |\overline{E}_1|^2 \qquad (8.25)$$

Example 8.1   Many experiments have been performed with carbon disulfide CS_2. The waves $\overline{E}_s, \overline{E}_3$ are polarized in the x-direction, and the waves $\overline{E}_1, \overline{E}_2$ in the y-direction. The relevant nonlinear coefficient is then χ_{1221}. From the intensities $I_1 = I_2 = 5 \times 10^{10}$ W/m^2, $n_{CS2} \approx 1.5$ at $\lambda = 1$ μm, and $|\chi_{\text{eff}}| = 22.8 \times 10^{-21}$ (m/V)^2 follows, [see Eq. (A.6)]:

$$|\overline{E}_1| = |\overline{E}_2| = \left(\frac{2Z_0 I_1}{n} \right)^{1/2} = 5 \times 10^6 \text{ V/m}, \quad \kappa = 1.2/\text{m}$$

Thus, with pump intensities on the order of MW/cm^2 on distances of 1 m, one can reach $\kappa L \approx 1$ and, therefore, $\tan \kappa L \approx 1.5$. In fact, phase conjugation has been observed already at much lower intensities because of self-focusing. ■

With a ruby laser at $\lambda = 0.694$ μm, one has obtained at a phase-conjugate mirror of 40-cm-long CS_2, an increase in intensity of $|\overline{E}_s(z = 0)/\overline{E}_3(z = 0)|^2$ of more the 200%.

Example 8.2   We consider the ABCD matrix formalism for phase conjugation. An incoming Gaussian beam [see Eq. (4.10)], can be written as

$$\Phi_i = \text{Re}\left[a(z) \cdot \exp j \left(\omega t - kz - \frac{kr^2}{2q_i} \right) \right], \qquad \frac{1}{q_i} = \frac{1}{R_i} - \frac{2j}{kw^2}$$

Phase conjugation is equivalent to time reversal and Φ_i is transformed into

$$\Phi_p = \text{Re}\left[a(z) \cdot e^{j(-\omega t - kz - kr^2/2q_i)} \right] = \text{Re}\left[a^*(z) \cdot e^{j(\omega t + kz + kr^2/2q_i^*)} \right]$$

$$= \text{Re}\left[a^*(z) \cdot e^{j(\omega t + kz - kr^2/2q_p)} \right]$$

with

$$\frac{1}{q_p} = -\frac{1}{q_i^*} = -\frac{1}{R_i} - \frac{2j}{kw^2}$$

So, an observer traveling with the beam will find the spot size w unchanged, but will see an opposite sign for the curvature of the wavefront: $R_p = -R_i$. Since $q_p = -q_i^*$, the ABCD transformation law of the form

$$q_p = q_p(q_i^*) = \frac{Aq_i^* + B}{Cq_i^* + D}$$

yields for phase conjugation

$$M_P = \begin{bmatrix} A & B \\ C & D \end{bmatrix} = \begin{bmatrix} 1 & 0 \\ 0 & -1 \end{bmatrix} \tag{8.26}$$

The important difference between the ABCD matrix for reflection at a plane mirror and that of the phase conjugation which looks identical, is as follows: With reflection, q_2 and q_1 are connected, with phase conjugation, q_2 and q_1^*. ∎

Applications of Phase Conjugation

1. *Nonlinear Laser Spectroscopy.* The investigation of the operation of a phase conjugator gives information about $\kappa = \kappa[\chi_{\text{eff}}^{(3)}]$ and, hence, on the third-order optical processes of the medium and the atomic system.

2. *Adaptive Optics.* This very important application means the restoration of the original state of a distorted and disturbed wave, even for strong aberrations; see Fig. 8.2. In conventional adaptive optics, however, one has to measure distortions of the wave front using sensors and, subsequently, small mirrors must be adjusted, for example, electromechanically. Such systems are expensive and slow.

3. *Optical Resonators with a Phase-Conjugating Mirror.* Even a phase variation of exactly zero can be produced. Distortions coming from the interior of the resonator can be removed. A phase-conjugating mirror can also contribute amplification. Such resonators are stable, independent of the radius of curvature of a conventional mirror, and independent of the length of the resonator.

4. *Real-Time Holography.* Whereas conventional holography is a two-step process (the first is the construction of the hologram on the film, the second step reconstruction of the object), with phase conjugation the process can proceed in a single step. Let $\left|\overline{E}_1\right| = \left|\overline{E}_2\right|$, $\kappa L \ll 1$ and $\left|\overline{E}_s(0)\right| \ll \left|\overline{E}_3(0)\right|$. Therefore, $|r| \ll 1$. According to Eq. (8.24), then

$$\overline{E}_s(0) = A \cdot \frac{\omega_0^2}{jk} \cdot L \cdot \overline{E}_3^* = \frac{\omega_0^2}{jk}\frac{\chi_{\text{eff}}^{(3)}}{2c^2} \cdot L \cdot \overline{E}_1\overline{E}_2\overline{E}_3^* = \frac{\omega_0}{2j}\sqrt{\frac{\mu_0}{\varepsilon}}\varepsilon_0 \cdot \chi_{\text{eff}}^{(3)}L \cdot \overline{E}_1\overline{E}_2\overline{E}_3^*$$

$$\tag{8.27}$$

Interpretation: $\overline{E}_1$ and $\overline{E}_3$ form a stationary holographic grating due to interference. $\overline{E}_s$ results through the Bragg reflection of $\overline{E}_2$ from this grating. The incident wave $\overline{E}_2$ deposits its amplitude and phase information in the holographic grating. The holographic object reconstruction yields the wave $\overline{E}_s$, which is identical to the phase-conjugated wave $\overline{E}_3^*$ up to a constant factor.

5. *Squeezed States in Quantum Optics.* In contrast to coherent light containing equal minimum field fluctuations (in a quantum theoretic sense) in the cophasal and quadrature component, in squeezed states one of these components acquires a smaller fluctuation at the expense of the other. These states arise in a resonator through a linear combination of the states of correlated conjugated pairs of photons, which are generated in degenerate FWM (from $\overline{E}_s$ and $\overline{E}_3$).

8.3 NONLINEAR OPTICAL SPECTROSCOPY

Lasers are ideal tools for high-resolving spectroscopy due to their extremely small linewidth. In conventional spectroscopy, inhomogeneous line broadening very often spoils this advantage. With the help of nonlinear optical spectroscopy, these difficulties can be avoided. One exploits the fact that susceptibility tensors, as they result explicitly from a classical consideration [see the eqs. (1.80) through (1.83)], or from a quantum theoretical treatment, have resonance denominators. The expression for $\chi_{\text{eff}}^{(3)}$ can show simple, double, or triple resonance poles, where the case of a simply resonant FWM can be tuned most easily. Therefore, it is possible to use a resonance effect independent of inhomogeneous broadening. Thus, in the method of *quantum beats*, beats occur after a resonant excitation of two neighboring energy levels due to a simultaneous spontaneous emission of light with two very closely lying frequencies (serving to investigate level splitting).

For *saturation spectroscopy*, only a group of molecules is selectively investigated with the same resonance frequency.

In *coherent anti-Stokes Raman scattering* (CARS), at first the material excitation $\omega_v = \omega_1 - \omega_2$ is coherently excited by the beats of two incident waves at ω_1, ω_2 and, subsequently, mixed with a wave at ω_1, resulting in a coherent output wave at the anti-Stokes frequency $\omega_a = \omega_v + \omega_1 = 2\omega_1 - \omega_2$. One must resort to the theory of the first case in Section 8.1; the signal intensity is proportional to $|\chi_{\text{eff}}^{(3)}|^2$. See Eq. (8.9).

In the *Raman-induced Kerr effect spectroscopy* (RIKES), the second case of Section 8.1 occurs, in which the output field coincides with one of the input fields. The signal shows loss or gain proportional to Im $\chi_{\text{eff}}^{(3)}$, see Eq. (8.12). Connected to the occurrence of Im $\chi_{\text{eff}}^{(3)}$ is then an induced birefringence (= Kerr effect).

8.4   THIRD HARMONIC GENERATION

We proceed similarly to the treatment of SHG in Chapter 4, substituting there $P^{(2)}(2f)$ by the expression Eq. (1.58). However, it is true that the third-order susceptibility tensor is considerably smaller in magnitude than the second-order tensor (in corresponding units), so a relatively high laser intensity is necessary.

In a crystal, laser intensity is often limited by optical damages in the crystal. Moreover, phase matching is not easily achievable. Therefore, the conversion efficiency for third harmonic generation (THG) in crystals is only about 10^{-6}.

A more efficient THG in crystals can be achieved in a two-step process. In the first step, the second harmonic wave at $2f$ is generated. In the second nonlinear crystal, this wave is then converted by a sum frequency mixing with f to $3f$. One obtains conversion efficiencies of 20%.

However, a good conversion efficiency can be obtained if gases are used as nonlinear media. The reasons for this are as follows: $\chi^{(3)}$ can grow up considerably in the neighborhood of resonances, and the limiting laser intensity in gases ($> \text{GW/cm}^2$) is much higher, as in solids (some 100 MW/cm^2). Thus, THG in Na-vapor, for example, is easily observable if $3f$ lies close to a resonance, for instance, with a Nd:YAG laser at $1.06 \, \mu\text{m}$.

The advantages of a gas medium for nonlinear optical mixing are (with permission from Shen, 1984):

1. One has a homogeneous medium over distances of more than 10 cm
2. With optimal focusing, the conversion efficiency can be increased.
3. Besides a high optical damage threshold, gases possess also a self-healing ability.
4. Atomic vapors are transparent for nearly all frequencies below the ionization level; in the range of the extreme UV and of the soft X-rays, they are actually the only usable nonlinear media.

Conversion is limited by intensity restrictions: In the resonant amplification of $\chi^{(3)}$, linear absorption also is increased a little; two-photon and multiphoton absorption (see the following section) represent an important limitation at high pump intensities; phase mismatch can be caused, for example, by a change in the refractive index (due to other lasing mechanisms); as already mentioned, a laser-induced destruction of the materials can limit the intensity.

Even higher harmonics have been generated experimentally in atomic vapor, for example, the fifth and seventh harmonic.

8.5   TWO-PHOTON ABSORPTION

Whereas for the Raman effect (Stokes line), $\omega_p - \omega_s = \omega_m$ holds, one has for *two-*

photon absorption

$$\omega_p + \omega_s = \omega_m \tag{8.28}$$

One arrives at an excited state of the molecule starting from the ground state by a simultaneous absorption of two photons with frequencies ω_p and ω_s. As for the Raman effect, we have here also a FWM. Neither ω_p nor ω_s should lie in the neighborhood of a transition frequency. The process is, indeed, much weaker than one-photon absorption, but can nevertheless easily be observed if laser light is used. The importance of two-photon absorption for spectroscopy lies in the fact that one has here other selection rules than for the one-photon process; that is, one can excite with two-photon absorption levels otherwise not attainable.

The theory proceeds similarly to the propagation of Raman light as discussed in Section 6.2. We only substitute in Eq. (6.34) ω_r (which could be equal to the Stokes frequency ω_s) by $-\omega_s$, and obtain then for the intensities the system of equations

$$\frac{\partial}{\partial z} I_p(z, f_p) = -\omega_p \cdot \gamma \cdot I_p(z, f_p) \cdot I_s(z, f_s)$$

$$\frac{\partial}{\partial z} I_s(z, f_s) = -\omega_s \cdot \gamma \cdot I_p(z, f_p) \cdot I_s(z, f_s) \tag{8.29}$$

where we have set $|\bar{E}|^2 \sim I$; see Eq. (A.6). Furthermore,

$$\gamma = \frac{2Z_0}{n_p n_s c} \cdot \text{Im } \chi$$

Instead of Eq. (6.35), we have here

$$\frac{I_p}{\omega_p} - \frac{I_s}{\omega_s} = \text{const} = K' = \frac{I_{p0}}{\omega_p} - \frac{I_{s0}}{\omega_s}, \qquad \frac{I_p - I_{p0}}{\omega_p} = \frac{I_s - I_{s0}}{\omega_s} \tag{8.30}$$

with the initial values

$$I_{p0} = I_p(z = 0), \qquad I_{s0} = I_s(z = 0)$$

$K' = $ const expresses the fact that as many photons at the frequency ω_p are annihilated as at the frequency ω_s. Using Eq. (8.30), one can integrate the system, Eq. (8.29), and get

$$I_p(z) = \frac{\omega_p K' I_{p0}}{I_{p0} - (\omega_p/\omega_s) I_{s0} \exp(-\gamma \omega_s \omega_p K' z)}$$

$$I_s(z) = \frac{\omega_s K' I_{s0}}{(\omega_s/\omega_p)I_{p0}\exp(+\gamma\omega_s\omega_p K'z) - I_{s0}} \tag{8.31}$$

If $I_{p0} \gg I_{s0}$, the decrease of I_p can be neglected and from Eq. (8.29) results

$$I_p(z) \approx I_{p0}$$
$$\tag{8.32}$$
$$I_s(z) \approx I_{s0} \cdot \exp(-\gamma\omega_s I_{p0}z)$$

We consider the special case of $\omega_p = \omega_s$. Then both Eqs. (8.29) coincide with $I_p = I_s$, Thus,

$$\frac{\partial}{\partial z}I_p = -\omega_p\gamma I_p^2$$

with the solution

$$I_p(z) = I_{p0} \cdot (1 + \omega_p\gamma I_{p0}z)^{-1}$$

or approximately for weak absorption (small γ)

$$I_p(z) \approx I_{p0} \cdot (1 - \omega_p\gamma I_{p0}z)$$

In measurements of two-photon absorption, one of the two input sources must be tunable. Since the tuning range of lasers unfortunately is small, very often an arc lamp in connection with a monochromator is used.

PROBLEMS

Section 8.1

8.1 Derive Eqs. (8.15).

8.2 Modify Eqs. (8.15) for the case of oscillation: $\kappa L \to \pi/2$.

Section 8.2

8.3 Equation (8.24) treats the case of phase-conjugated reflection. Derive now the expression for transmission through the phase-conjugated mirror.

8.4 Consider an optical resonator in which the reflection at a conventional mirror on the left with a radius of curvature R is given by the matrix

$$M_M = \begin{pmatrix} 1 & 0 \\ -\dfrac{2}{R} & 1 \end{pmatrix}$$

At the right end, the resonator is bounded by a phase-conjugated mirror with a matrix M_P; see Eq. (8.26). The propagation through the resonator from left to right is described by M_L, and from right to left by M_R.

(a) Calculate the matrix M_1 for one round-trip, using the relation $M_R M_P = M_P (M_L)^{-1}$ following from reciprocity.

(b) Show that even the phase will be restored after two round-trips. *Hint*: Calculate $M_2 = M_1^2$.

Section 8.5

8.5 Derive the Eqs. (8.31).

BIBLIOGRAPHY

G. P. Agrawal. *Nonlinear Fiber Optics.* Wiley, New York, 1984.

P. A. Fisher (ed.). *Optical Phase Conjugation.* Academic Press, New York, 1983.

B. E. A. Saleh and M. C. Teich. *Fundamentals of Photonics.* Wiley, New York, 1991.

M. Schubert and B. Wilhelmi. *Nonlinear Optics and Quantum Electronics.* Wiley, New York, 1986.

Y. R. Shen. *The Principles of Nonlinear Optics.* Wiley, New York, 1984.

A. Yariv. *Optical Electronics.* Holt, Rinehart and Winston, New York, 1985.

CHAPTER NINE

Propagation of Light Pulses

With this chapter, we begin the treatment of pulse propagation. In the first section, we describe the propagation of light pulses in dispersive, but linear media, that is, the refractive index is assumed to be independent of the electric field E. In the second section, propagation in nonlinear media is discussed, which leads to self-phase modulation and solitons, the discussion of which is delayed until Chapter 10. Modulation instability can break a pulse into a series of short pulses due to the simultaneous occurrence of dispersion and nonlinearity.

9.1 LIGHT PULSES IN DISPERSIVE LINEAR MEDIA

9.1.1 The Initial State at z = 0

As a model, we consider an optical input pulse with a Gaussian envelope (analytic signal), since then all integrals (e.g., Fourier integrals) can be calculated easily. The results should not differ very much from those valid for a pulse that is only approximately Gaussian. So we use

$$\widetilde{E}(z=0,t) = E_0 \cdot e^{-a_0 t^2} e^{j\phi(t)} = E_0 \cdot e^{-a_0 t^2} e^{j(\omega_0 t + b_0 t^2)} = E_0 \cdot e^{-\Gamma_0 t^2} e^{j\omega_0 t} \quad (9.1)$$

as an input pulse with

$$E_0 = \widetilde{E}(0,0), \quad a_0 > 0, \quad -\infty < t < \infty, \quad \Gamma_0 = a_0 - jb_0, \quad \phi(t) = \omega_0 t + b_0 t^2$$

and with an intensity

$$I(0,t) = I_0 \cdot e^{-2a_0 t^2} = I_0 \cdot e^{-\ln 2 \cdot (2t/\Delta t_0)^2} = I_0 \cdot e^{-t^2/(2\sigma_{t_0}^2)}$$

$$I_0 := \frac{n}{2Z_0} \cdot E_0^2 = \frac{P_0}{A} \quad (9.2)$$

See Eq. (A.6). I_0, $\mathcal{P}_0$ are the peak values of intensity and power at $z = 0$, $t = 0$, respectively; A is the cross-sectional area of the medium. The *pulse width* (in the intensity) of this Gaussian pulse can be characterized by the full width at half maximum (FWHM)

$$\Delta t_0 = \sqrt{\frac{2 \cdot \ln 2}{a_0}} = \sqrt{\frac{2 \cdot \ln 2}{\text{Re } \Gamma_0}} \tag{9.3}$$

or by the standard deviation

$$\sigma_{t_0} = \frac{1}{\sqrt{4a_0}} = \frac{1}{\sqrt{4 \cdot \text{Re } \Gamma_0}} = \Delta t_0 / \sqrt{8 \cdot \ln 2} \tag{9.4}$$

The total phase is

$$\phi = \omega_0 t + b_0 t^2 \tag{9.5}$$

so we find as the instantaneous frequency

$$\omega(t) := \frac{d\phi}{dt} = \omega_0 + 2b_0 t \tag{9.6}$$

b_0 is the chirp parameter responsible for the time dependence of ω (*linear chirp*; $b_0 > 0$: up chirp, $b_0 < 0$: down chirp).

The Fourier transform of Eq. (9.1) is

$$\tilde{E}(0, \omega) = \int_{-\infty}^{\infty} \tilde{E}(0, t) \cdot e^{-j\omega t} dt = E_0 \cdot \int_{-\infty}^{\infty} e^{-\Gamma_0 t^2 + j(\omega_0 - \omega)t} dt$$

$$= \tilde{E}(0, \omega_0) \exp\left[-\frac{(\omega - \omega_0)^2}{4\Gamma_0}\right]$$

$$= \tilde{E}(0, \omega_0) \cdot \exp\left[-\frac{a_0(\omega - \omega_0)^2}{4(a_0^2 + b_0^2)} - j\frac{b_0(\omega - \omega_0)^2}{4(a_0^2 + b_0^2)}\right] \tag{9.7}$$

$\tilde{E}(0, \omega)$ is a function of $\omega - \omega_0$, $\tilde{E}(0, \omega_0) = E_0\sqrt{\pi/\Gamma_0}$. (In Chapter 9, frequency dependence will be expressed in the angular frequency.) Thus, according to Eq. (A.6), as power spectral density at $z = 0$

$$I(z = 0, \omega) = I_{0\omega} \cdot \exp\left[-\text{Re}\left(\frac{1}{\Gamma_0}\right) \cdot \frac{(\omega - \omega_0)^2}{2}\right] - I_{0\omega} \cdot \exp\left[\frac{a_0(\omega - \omega_0)^2}{2(a_0^2 + b_0^2)}\right] \tag{9.8}$$

$$I_{0\omega} := \left[\frac{n}{(2Z_0)} \right] \cdot |\tilde{E}(0, \omega_0)|^2 = \frac{I_0 \pi}{|\Gamma_0|}$$

Equation (9.8) is a Gaussian distribution too, centered at the angular frequency ω_0. From the intensity, we again read off the *pulse bandwidth*: $\Delta\omega$ = FWHM and σ_ω = standard deviation:

$$\Delta\omega = \sqrt{8 \cdot \ln 2 \cdot a_0 \left(1 + \frac{b_0^2}{a_0^2} \right)} = \sqrt{\frac{8 \cdot \ln 2}{\text{Re}\,(1/\Gamma_0)}} = \sqrt{\frac{8 \cdot \ln 2}{\text{Re}\,\Gamma_0}} \cdot |\Gamma_0| = 2\pi \Delta f \quad (9.9)$$

$$\sigma_\omega = \sqrt{a_0 \left(1 + \frac{b_0^2}{a_0^2} \right)} = \frac{1}{\sqrt{\text{Re}\,(1/\Gamma_0)}} = \frac{|\Gamma_0|}{\sqrt{\text{Re}\,\Gamma_0}} = \frac{\Delta\omega}{\sqrt{8 \ln 2}} = 2\pi \sigma_f \quad (9.10)$$

thus,

$$\frac{\Delta\omega(b_0 \neq 0)}{\Delta\omega(b_0 = 0)} = \frac{\sigma_\omega(b_0 \neq 0)}{\sigma_\omega(b_0 = 0)} = \sqrt{1 + \left(\frac{b_0}{a_0} \right)^2}$$

Note that

$$\sigma_{t_0} \cdot \sigma_\omega = \frac{|\Gamma_0|}{(2 \cdot \text{Re}\,\Gamma_0)} = \frac{1}{2} \cdot \sqrt{1 + \left(\frac{b_0}{a_0} \right)^2} \geq \frac{1}{2} \quad (9.11)$$

Pulses with $\sigma_{t_0} \cdot \sigma_\omega = 1/2$ are referred to as *transform-limited pulses*.

9.1.2 Passage through the Medium ($z > 0$)

A nondispersive medium has a refractive index $n(\omega)$ = const and a constant of propagation $\beta(\omega) = (\omega/c)n$ (only the vacuum is strictly nondispersive), whereas a dispersive medium with the propagation constant

$$\beta(\omega) = k_0 n(\omega) = \frac{\omega}{c} \cdot n(\omega) = \frac{2\pi}{\lambda} \cdot n(\lambda) \quad (9.12)$$

can be characterized by an expansion (up to the quadratic term)

$$\beta(\omega) = \beta(\omega_0) + \beta'(\omega_0) \cdot (\omega - \omega_0) + \frac{1}{2!} \cdot \beta''(\omega_0) \cdot (\omega - \omega_0)^2 \quad (9.13)$$

(chromatic dispersion). Here,

$$\beta'(\omega_0) = \beta_0' = \frac{d\beta}{d\omega} = \frac{1}{v_g} = \frac{n_g}{c} = \left(n + \omega \cdot \frac{dn}{d\omega} \right) \frac{1}{c}$$

$$\beta''(\omega_0) = \beta_0'' = \frac{d^2\beta}{d\omega^2} = \frac{d}{d\omega}\left[\frac{1}{v_g(\omega)}\right]$$

$$= -\frac{1}{v_g^2}\frac{dv_g}{d\omega} = \frac{\lambda^3}{2\pi c^2}\frac{d^2n}{d\lambda^2} = -\frac{\lambda^2}{2\pi c}\cdot D \qquad (9.14)$$

All expressions are evaluated at $\omega = \omega_0$; $\beta_0 = \beta(\omega_0)$; n_g is the group index and

$$D = \frac{d\beta'}{d\lambda} = -\frac{2\pi c}{\lambda^2}\beta_0'' = -\frac{\lambda}{c}\frac{d^2n}{d\lambda^2}$$

the dispersion parameter. If $\beta''(\omega_0) \neq 0$, the group velocity v_g is dependent on ω, that is,

$$v_g(\omega) = v_g(\omega_0) - \beta_0'' \cdot v_g^2(\omega_0) \cdot (\omega - \omega_0) \qquad (9.15)$$

and the medium is dispersive. One has a $\{{}^{\text{positive}}_{\text{negative}}\}$ group-velocity dispersion $\beta_0'' \gtrless 0$, if $d^2n/d\lambda^2 \gtrless 0$, or if $dv_g/d\omega \lessgtr 0$. Glass has a zero of dispersion at $\lambda = \lambda_D \approx 1.3\,\mu\text{m}$. For $\lambda < \lambda_D$, one has $\beta_0'' > 0$ (regime of normal dispersion); for $\lambda > \lambda_D$, one has $\beta_0'' < 0$ (regime of anomalous dispersion). [At the "zero of dispersion" higher terms in the expansion Eq. (9.13) should be considered.] Higher terms in the expansion Eq. (9.13) can be neglected (for $\lambda \neq \lambda_D$) if the bandwidth $\Delta\omega$ is much smaller than the carrier frequency ω_0 (this is consistent with the assumption of a quasimonochromatic wave). However, for pulse widths ≤ 0.1 ps, $\Delta\omega$ comes close to ω_0 and the theory must be changed.

At first, we consider here the linear propagation effects caused by a linear frequency response of the systems. For the wave at position z, we find in the frequency domain

$$\widetilde{E}(z,\omega) = \widetilde{E}(0,\omega) \cdot e^{-j\beta(\omega)z}$$

$$= \widetilde{E}(0,\omega) \cdot e^{-jz[\beta_0 + \beta_0'(\omega - \omega_0) + \beta_0''(\omega - \omega_0)^2/2]}$$

$$=: e^{-j\beta_0 z} \cdot \widetilde{E}(z, \omega - \omega_0) \qquad (9.16)$$

$[\widetilde{E}(z, \omega - \omega_0)$ is the slowly varying envelope], or

$$|\widetilde{E}(z,\omega)|^2 = |\widetilde{E}(0,\omega)|^2, \qquad I(z,\omega) = I(0,\omega) \qquad (9.17)$$

that is, the bandwidth of the pulses [$\Delta\omega$, σ_ω, of Eq. (9.9) and Eq. (9.10)] does not change in the linear case; it is independent of z and would not vary even if the expansion in Eq. (9.13) is extended. So the expressions for $\Delta\omega$, σ_ω hold for $z \geq 0$ too.

However, with the abbreviations

$$\frac{1}{\Gamma(z)} = \frac{1}{\Gamma_0} + 2j\beta_0'' z \tag{9.18}$$

and

$$\Gamma(z) = a(z) - jb(z), \quad \Gamma_0 = \Gamma(0) = a_0 - jb_0 \tag{9.19}$$

we get for the field in the time domain (Fourier back transform)

$$
\begin{aligned}
\tilde{E}(z,t) &= \int_{-\infty}^{\infty} \tilde{E}(z,\omega) \cdot e^{j\omega t} df \\
&= \tilde{E}(0,\omega_0) e^{j(\omega_0 t - \beta_0 z)} \int_{-\infty}^{\infty} \exp\left[-\frac{a_0(\omega - \omega_0)^2}{4(a_0^2 + b_0^2)} - j\frac{b_0(\omega - \omega_0)^2}{4(a_0^2 + b_0^2)} \right. \\
&\qquad \left. -j\frac{\beta_0'' z}{2}(\omega - \omega_0)^2 - jz\beta_0'(\omega - \omega_0) + j(\omega - \omega_0)t \right] df \\
&= E_0 \cdot \sqrt{\frac{\Gamma(z)}{\Gamma(0)}} \cdot \exp[j\omega_0(t - t_\varphi)] \cdot \exp[-\Gamma(z) \cdot (t - t_g)^2] \tag{9.20}
\end{aligned}
$$

with a phase delay

$$t_\varphi = \frac{z}{v} = \frac{z\beta_0}{\omega_0}$$

and group delay

$$t_g = z\beta_0' = \frac{z}{v_g}$$

For the phase at $z \neq 0$ and the instantaneous frequency, we obtain therefore

$$\phi(z,t) = \omega_0(t - t_\varphi) + b(z) \cdot (t - t_g)^2 \tag{9.21}$$

$$\omega(z,t) = \frac{d}{dt}\phi(z,t) = \omega_0 + 2b(z) \cdot (t - t_g) = [\omega_0 - 2b(z) \cdot t_g] + 2b(z) \cdot t \tag{9.22}$$

with a z-dependent chirp, given by $b(z)$.

From Eqs. (9.18) and (9.19) result

$$
\begin{aligned}
a(z) &= \frac{a_0}{(1 + 2\beta_0'' z b_0)^2 + (2\beta_0'' z a_0)^2} \\
b(z) &= \frac{b_0(1 + 2\beta_0'' z b_0) + 2\beta_0'' z a_0^2}{(1 + 2\beta_0'' z b_0)^2 + (2\beta_0'' z a_0)^2}
\end{aligned}
\tag{9.23}
$$

From Eq. (9.20), we get for the intensity

$$I(z,t) = I(z,t_g) \cdot \exp[-2a(z) \cdot (t - t_g)^2], \qquad I(z,t_g) = I_0 \left| \frac{\Gamma(z)}{\Gamma_0} \right| \qquad (9.24)$$

This yields after a distance z a pulse width of Δt (FWHM) and σ_t (standard deviation), respectively, of

$$\Delta t = \sqrt{\frac{2 \cdot \ln 2}{\mathrm{Re}\, \Gamma(z)}} = \sqrt{\frac{2 \cdot \ln 2}{a(z)}} = \sqrt{\frac{2 \cdot \ln 2}{a_0}} \cdot \sqrt{(1 + 2\beta_0'' z b_0)^2 + (2\beta_0'' z a_0)^2}$$

$$= \Delta t_0 \cdot \sqrt{(1 + 2\beta_0'' z b_0)^2 + (2\beta_0'' z a_0)^2}$$

$$\sigma_t = \sqrt{\frac{1}{4 \cdot \mathrm{Re}\, \Gamma(z)}} = \frac{1}{\sqrt{4 \cdot a_0}} \cdot \sqrt{(1 + 2\beta_0'' z b_0)^2 + (2\beta_0'' z a_0)^2} \qquad (9.25)$$

$$= \sigma_{t_0} \cdot \sqrt{(1 + 2\beta_0'' z b_0)^2 + (2\beta_0'' z a_0)^2} = \frac{\Delta t}{\sqrt{8 \ln 2}}$$

9.1.3 Differential Equations

The elimination of **H** from the Maxwell equations (2.1) gives (take $\sigma = 0$)

$$\nabla \times \nabla \times \mathbf{E}(z,t) = -\mu_0 \ddot{\mathbf{D}}(z,t)$$

Let $\mathbf{E}(z,t) = E(z,t)\mathbf{e}_x$; then we have $\nabla \cdot \mathbf{E} = \mathbf{0}$, and we find

$$\Delta \mathbf{E}(z,t) = \partial_z^2 \mathbf{E}(z,t) = \mu_0 \ddot{\mathbf{D}}(z,t)$$

or, after a Fourier transform

$$\partial_z^2 E(z,\omega) + \mu_0 \omega^2 D(z,\omega) = 0$$

In the frequency domain, we set, according to Eq. (1.3),

$$D(z,\omega) = \varepsilon(\omega, E) \cdot E(z,\omega)$$

and obtain

$$\partial_z^2 E(z,\omega) + \beta^2(\omega, E) \cdot E(z,\omega) = 0 \qquad (9.26)$$

since

$$\omega^2 \mu_0 \varepsilon(\omega, E) = \omega^2 \mu_0 \varepsilon_0 \varepsilon_r(\omega, E) = k_0^2 n^2(\omega, E) =: \beta^2(\omega, E)$$

Owing to the frequency dependence of n^2, dispersion is taken into account; nonlinearity is described by the dependence on E. We use now

$$E(z,t) = \text{Re}\widetilde{E}(z,t) = \text{Re}[\bar{E}(z,t)e^{j(\omega_0 t - \beta_0 z)}] = \frac{1}{2}\left[\bar{E}(z,t)e^{j(\omega_0 t - \beta_0 z)} + \text{c.c.}\right] \quad (9.27)$$

and $\bar{E}(z,t)$ is only weakly dependent on z and t. A Fourier transform yields

$$
\begin{aligned}
E(z,\omega) &= \int E(z,t)e^{-j\omega t}\,dt \\
&= \frac{1}{2}e^{-j\beta_0 z}\int \bar{E}(z,t)e^{-j(\omega-\omega_0)t}\,dt + \frac{1}{2}e^{j\beta_0 z}\left[\int \bar{E}(z,t)e^{+j(\omega+\omega_0)t}\,dt\right]^* \\
&= \frac{1}{2}e^{-j\beta_0 z}\bar{E}(z,\omega-\omega_0) + \frac{1}{2}e^{j\beta_0 z}\bar{E}^*(z,-\omega-\omega_0)
\end{aligned}
$$

If $\bar{E}(z,t) = \bar{E}(z)$, that is, constant in time, then

$$\bar{E}(z,\omega-\omega_0) = \bar{E}(z)\cdot\delta(f-f_0), \qquad \bar{E}^*(z,-\omega-\omega_0) = \bar{E}^*(z)\cdot\delta(f+f_0)$$

and this second part does not yield a contribution due to $f \neq -f_0$. For a weak time dependence, this part can be neglected at least approximately. After insertion in Eq. (9.26) and in the approximation of slowly varying amplitudes, that is, $|\bar{E}''| \ll |\beta_0\bar{E}'|$, one finally finds

$$-2j\beta_0\partial_z\bar{E}(z,\omega-\omega_0) + (\beta^2 - \beta_0^2)\bar{E}(z,\omega-\omega_0) = 0$$

or

$$\partial_z\bar{E}(z,\omega-\omega_0) + j(\beta - \beta_0)\bar{E}(z,\omega-\omega_0) = 0 \quad (9.28)$$

if we set approximately

$$\beta^2 - \beta_0^2 = (\beta + \beta_0)(\beta - \beta_0) \approx 2\beta_0(\beta - \beta_0)$$

We consider in Section 9.1 only the linear case: ε, n, β independent of $\bar{E}$, and take the expression Eq. (9.13) for $\beta - \beta_0$. This results in

$$\partial_z\bar{E}(z,\omega-\omega_0) + j\left[\beta_0'\,(\omega-\omega_0) + \frac{\beta_0''}{2}(\omega-\omega_0)^2\right]\bar{E}(z,\omega-\omega_0) = 0$$

We multiply with $\exp(j\omega t)$ and integrate over f from $-\infty$ to ∞ (Fourier back transform). In the time domain, this yields the differential equation

$$j\frac{\partial}{\partial z}\bar{E}(z,t) + j\beta_0'\frac{\partial}{\partial t}\bar{E}(z,t) + \frac{\beta_0''}{2}\frac{\partial^2}{\partial t^2}\bar{E}(z,t) = 0 \quad (9.29)$$

or, with the new independent variables τ, ζ,

$$\begin{cases} \tau = t - \dfrac{z}{v_g} \\ \zeta = z \end{cases} \quad \text{or} \quad \begin{cases} t = \tau + \dfrac{\zeta}{v_g} \\ z = \zeta \end{cases} \tag{9.30}$$

$$j \frac{\partial}{\partial \zeta} \bar{E}(\zeta, \tau) = -\frac{1}{2} \cdot \beta_0'' \cdot \frac{\partial^2}{\partial \tau^2} \bar{E}(\zeta, \tau) \tag{9.31}$$

This parabolic differential equation has the form of a (linear) Schrödinger equation.

With the ansatz

$$\bar{E}(z, t) = A \cdot \exp[j(\Delta\omega \cdot t - \Delta\beta \cdot z)]$$

and

$$\Delta\omega = \omega - \omega_0, \qquad \Delta\beta = \beta - \beta_0$$

again we get exactly from Eq. (9.29) the dispersion relation Eq. (9.13). This expression for $\bar{E}(z, t)$ substituted in Eq. (9.27) yields

$$\tilde{E} = A \exp j(\omega t - \beta z)$$

for the complex analytic signal.

Since with Eq. (9.18) also

$$\frac{d}{dz} \frac{1}{\Gamma(z)} = -\frac{\Gamma'(z)}{\Gamma^2(z)} = 2j\beta_0'' \tag{9.32}$$

is valid, we find the differential equations

$$\Gamma'(z) = -2j\beta_0'' \Gamma^2(z), \quad \begin{cases} \dfrac{da}{dz} = -4\beta_0'' a(z) b(z) \\ \dfrac{db}{dz} = 2\beta_0'' [a^2(z) - b^2(z)] \end{cases} \tag{9.33}$$

Example 9.1 Let us consider an alternative derivation of the differential equation from the dispersion relation and substitution rule. The differentiation of $\tilde{E}(z, \omega - \omega_0)$ from Eq. (9.16) with respect to z and a Fourier transform into the time domain yields immediately Eq. (9.29). Using the ansatz $\tilde{E} = A \cdot \exp j(\omega t - \beta z)$ in Eq. (9.27) and substituting then $\bar{E}$ in Eq. (9.29) give the dispersion relation Eq. (9.13). Vice versa, one gets the differential equation (9.29) from the dispersion relation by the substitution $\partial_z \leftrightarrow -j\Delta\beta, \partial_t \leftrightarrow j\Delta\omega$, if applied to $\bar{E}(z, t)$. ∎

9.1.4   Pulse Broadening and Compression

To begin with the case without dispersion ($\beta_0'' = 0$), it is clear that not only the bandwidth remains constant when propagating, but also the pulse width, in Eq. (9.25). Then, according to Eq. (9.23), $a(z) = a_0$, $b(z) = b_0$, and $\Gamma(z) = \Gamma_0$. The pulse, Eq. (9.20), maintains its initial shape; there is only a time shift in the phase, Eq. (9.21), and a shift in the maximum of the pulse.

Now we come to $\beta_0'' \neq 0$. As Eq. (9.25) shows, one has always a pulse broadening if $\beta_0'' b_0 > 0$, whereas for $\beta_0'' b_0 < 0$ and small values of z and a_0, we have at first a pulse compression that turns into pulse broadening for larger z values. If pulses propagate in monomode-glass fibers with $\beta_0'' b_0 < 0$, they can propagate very long before considerable broadening begins.

Δt is minimal for $a(z) = \mathrm{Re}\ \Gamma(z)$ maximal, or $b(z) = 0$ [then $\Gamma(z)$ and $1/\Gamma(z)$ are real]; this is the point D in Fig 9.1. The figures hold for $b_0 < 0$; the arrows show the movement of the system point in the complex $1/\Gamma(z)$ and the complex $\Gamma(z)$ plane, respectively, for $z \to \infty$ in the two cases $\beta_0'' \gtrless 0$. From Fig. 9.1, one reads directly (this can also be found after a simple calculation)

$$a_{\max} = a_0 \left[1 + \left(\frac{b_0}{a_0} \right)^2 \right] \approx \frac{b_0^2}{a_0}, \qquad b_{\mathrm{opt}} = b(z_{\mathrm{opt}}) = 0$$

(The approximation holds for $b_0^2 \gg a_0^2$). Then, there is no longer a chirp, and the minimum pulse width is

$$\Delta t_{\min} = \sqrt{\frac{2 \cdot \ln 2}{a_{\max}}} = \sqrt{\frac{2 \cdot \ln 2}{a_0}} \left[1 + \left(\frac{b_0}{a_0} \right)^2 \right]^{-1/2} \approx \Delta t_0 \cdot \left| \frac{a_0}{b_0} \right| \ll \Delta t_0 \quad (9.34)$$

for $b_0^2 \gg a_0^2$. Using $b(z = z_{\mathrm{opt}}) = b_{\mathrm{opt}} = 0$ in Eq. (9.23) shows that the optimal

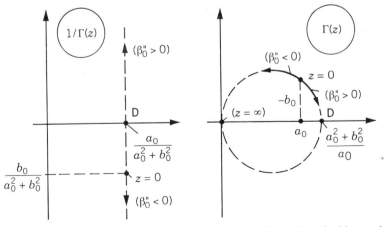

FIGURE 9.1   Trajectories in two complex planes for $b_0 < 0$. (Reproduced with permission from A. E. Siegman, *Lasers*, University Science Books, Mill Valley, CA., 1986.)

value of z is given by

$$z_{\text{opt}} = -\frac{b_0}{a_0^2 + b_0^2} \cdot \frac{1}{2\beta_0''} \approx -\frac{1}{2\beta_0'' b_0} > 0 \qquad (9.35)$$

that is, $\beta_0'' b_0 < 0$ as already mentioned. Given an initial chirp $b_0 \neq 0$, that is an initial time-bandwidth product $> 1/2$, the minimum value for this product occurs after a distance z_{opt} [use Eqs. (9.10) and (9.25)]:

$$\sigma_{t,\text{min}} \cdot \sigma_\omega = \frac{1}{2}$$

Explanation in the Time Domain We can think of the pulse as being divided into a number of small segments. At time $t = 0$, let the central portion start at $z = 0$ with an instantaneous frequency $\omega(t = 0) = \omega_0$ and a group velocity $v_g = v_g(\omega_0) = v_{g0}$. After a distance z, this part has a group delay $t_{g0} = z/v_{g0}$. Let another part start at time t_1 ($t_1 > 0$: later; $t_1 < 0$: earlier) with an instantaneous frequency $\omega_1 \equiv \omega(t_1) = \omega_0 + 2b_0 t_1$, [see Eq. (9.6)] and a group velocity $v_g(\omega_1) = v_{g0} - \beta_0'' v_{g0}^2 \cdot (\omega_1 - \omega_0) = v_{g0} - \beta_0'' v_{g0}^2 \cdot 2b_0 t_1$, [see Eq. (9.15)]. After a distance z, this part has a group delay of $t_{g1} = z/v_{g1} \approx (z/v_{g0})$ $(1 + 2\beta_0'' v_{g0} b_0 t_1)$. Both parts of the pulse arrive at position z exactly at the same time if $t_1 + t_{g1} = t_{g0}$, from where we find again $z \approx -1/(2\beta_0'' b_0)$; see Eq. (9.35). This pulse compression causes the leading edge of the pulse to run a little slower and the trailing edge of the pulse a little faster than the central part.

Explanation in the Frequency Domain If a medium has $\beta_0'' < 0$, then according to Eq. (9.14) $dv_g/d\omega > 0$, and the group velocity grows with increasing frequency (and vice versa). For $t < 0$ and $b_0 > 0$, the frequency is reduced according to Eq. (9.6), that is, the group velocity of this pulse portion decreases too: The leading edge of the pulse propagates more slowly, similarly, the trailing edge propagates faster. Altogether a pulse compression occurs. Subsequently, the trailing edge overtakes the leading edge and becomes itself the leading edge, with the previously leading edge becoming the trailing edge. Now, pulse broadening starts. Exactly the same holds for $\beta_0'' > 0, b_0 < 0$.

If initially there is no chirp, $b_0 = 0$ (we start at point D in Fig. 9.1), then in the course of propagation a chirp appears, [see Eq. (9.23)]:

$$b(z) = \frac{2\beta_0'' z a_0^2}{1 + (2\beta_0'' z a_0)^2} \neq 0 \qquad (9.36)$$

The chirp parameter $b(z)$ is positive for $\beta_0'' > 0$. The pulse can only be broadened now, since

$$a(z) = \frac{a_0}{1 + (2\beta_0'' z a_0)^2} \leq a_0 \qquad (9.37)$$

$a(z)$ is decreasing with increasing z; therefore Δt increases [see Eq. (9.25)], or

$$(\Delta t)^2 = \frac{2 \cdot \ln 2}{a(z)} = (\Delta t_0)^2 \cdot \left[1 + \left(\frac{z}{z_D} \right)^2 \right] \tag{9.38}$$

with a dispersion length

$$z_D = \frac{1}{2 \cdot |\beta_0''| \cdot a_0} = \frac{(\Delta t_0)^2}{4 \cdot \ln 2 \cdot |\beta_0''|} = 2 \frac{\sigma_{t_0}^2}{|\beta_0''|} \tag{9.39}$$

Δt has been increased by a factor of $\sqrt{2}$ after a distance of $z = z_D$

We consider now pulses in monomode-glass fibers with $b_0 = a_0 > 0, \beta_0'' < 0$. After a distance of $z = z_{\mathrm{opt}}$, the pulse has a minimum width of $\Delta t(z_{\mathrm{opt}}) = \Delta t_{\min} = \Delta t_0 / \sqrt{2}$ with $b(z_{\mathrm{opt}}) = b_{\mathrm{opt}} = 0$. During the remaining distance of length z_D, a chirp is built up and the pulse width increases again by a factor of $\sqrt{2}$ to $\Delta t(z_{\mathrm{opt}} + z_D) = \sqrt{2} \, \Delta t(z_{\mathrm{opt}}) = \Delta t_0$, that is, to the original width. It is only then that the actual pulse broadening starts.

A method of pulse compression known in microwave technics is chirp radar. In optics, a pulse with normal dispersion ($\beta_0'' > 0$) and an up chirp ($b_0 > 0$) hits a pair of parallel reflection gratings, as shown in Fig. 9.2; then the blue part of the spectrum that occurs at the trailing edge ($t > 0$) of the pulse belongs to a smaller angle of diffraction than the red part, that is, the blue part covers a shorter path and arrives earlier: The pulse is compressed. We show this now explicitly. The different frequency components are diffracted from the first grating in different directions, and we have for the diffraction angle θ_R, with a grating constant Λ,

$$\sin \theta_R = \frac{2\pi c}{\omega \cdot \Lambda} - \sin \theta_I = \frac{\lambda}{\Lambda} - \sin \theta_I \tag{9.40}$$

(θ_I = angle of incidence). A corresponding relation holds at the second grating. So a frequency-dependent group delay $t_g(\omega)$ exists after a distance $L(\omega) = L_1(\omega) + L_2(\omega)$ through the grating pair and a phase shift $\hat{\phi}(\omega)$, defined

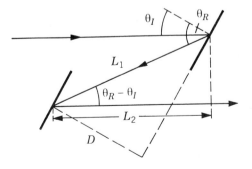

FIGURE 9.2 Pulse compression at a pair of parallel reflection gratings.

in free space by

$$\cos(\omega t + \hat{\phi}) = \cos[\omega t - \beta(\omega)z] = \cos \omega \left(t - \frac{z}{c} \right) \qquad (9.41)$$

Thus,

$$t_g(\omega) = \frac{L(\omega)}{v_g} = L\beta' = L \cdot \frac{d\beta}{d\omega} = \frac{L}{c} = -\frac{d\hat{\phi}(\omega)}{d\omega} \qquad (9.42)$$

From the geometry of the arrangement follows:

$$L(\omega) = L_1 + L_2 = \frac{D \cdot (1 + L_2/L_1)}{D \cdot 1/L_1} = D \cdot \frac{1 + \cos(\theta_R - \theta_I)}{\cos \theta_R} \qquad (9.43)$$

Finally, one expands $\hat{\phi}(\omega)$ about a central frequency ω_0:

$$\hat{\phi}(\omega) = \hat{\phi}_0 + t_c \cdot (\omega - \omega_0) + a_c \cdot (\omega - \omega_0)^2 + \ldots \qquad (9.44)$$

with constants t_c, a_c. We now eliminate θ_I in $L(\omega)$ using Eq. (9.40) (written down for ω_0 and θ_{R0}) and expand $L(\omega)$ with respect to ω about ω_0. Moreover, let us assume that only small angles θ_{R0} occur: $\sin \theta_{R0} \ll \lambda_0/\Lambda$. Comparing with Eqs. (9.42) and Eq. (9.44) then yields

$$a_c = \frac{2\pi^2 \cdot cL_1(\omega_0)}{\omega_0^3 \cdot \Lambda^2 \cdot \cos^2 \theta_{R0}} > 0 \qquad (9.45)$$

with a mean distance between both gratings of $L_1(\omega_0) = D/\cos \theta_{R0}$. Thus, the third term on the right-hand side of Eq. (9.44) is positive. Comparing this result with the phase of a medium with dispersion:

$$\cos(\omega t + \hat{\phi}) = \cos\{\omega t + \hat{\phi}_0 + t_c \cdot (\omega - \omega_0) + a_c \cdot (\omega - \omega_0)^2 + \ldots\}$$

$$= \cos\left\{ \omega t - \left[\beta_0 + \beta_0'(\omega - \omega_0) + \frac{\beta_0''}{2} \cdot (\omega - \omega_0)^2 \right] \cdot z \right\}$$

one finds that $a_c > 0$ just corresponds to $\beta_0'' < 0$.

Thus, with this grating arrangement, the sign of β_0'' has become negative. As already described in detail, this causes a pulse compression ($b_0 > 0$). One has obtained experimentally pulse lengths of 6 fs at $\lambda = 620$ nm and a compression factor of 5000 at $\lambda = 1.32 \, \mu$m.

Usually, the pulse leaving the grating pair will be reflected back, so the effect of dispersion will be doubled and the length $L_1(\omega_0)$ can be halved. Moreover, this helps to avoid the diffraction losses from the gratings.

Instead of a grating pair, one can also use a pair of prisms or a Gires–

Tournois interferometer (Fabry–Perot interferometer in reflection, with a frequency-dependent phase shift). Common to these arrangements is always a negative value for β_0''.

9.2 LIGHT PULSES IN NONLINEAR MEDIA

9.2.1 Differential Equations and Classification

The pulse broadening described in Eqs. (9.36) through (9.39) at $b_0 = 0$ due to dispersion can be canceled in a nonlinear medium possessing a refractive index of the form

$$n_{\text{tot}} = n + \delta n, \qquad \delta n = n_2 \cdot |\tilde{E}|^2 = n_2' I = n_2'' \mathcal{P}$$

$$n_2 = n_2' \cdot \frac{n}{2Z_0}, \qquad n_2' = n_2'' A \tag{9.46}$$

[see Eq. (7.8), power $\mathcal{P} = IA$; for quartz glass one has, e.g., $n_2 \approx 1.2 \times 10^{-22}$ m^2/V^2, $n_2' = 3 \dots 6 \times 10^{-20}$ m^2/W]. The effect of self-phase modulation, the counterpart in time of spatial self-focusing, counteracts, as we shall describe later, with dispersive pulse broadening. The propagation of an optical pulse through a dispersive nonlinear medium can be described by a nonlinear Schrödinger equation.

To begin with, we substitute $\beta(\omega_0) = \beta_0$ in Eq. (9.13) with

$$\beta_0 = \frac{\omega_0}{c} \cdot n_{\text{tot}}(\omega_0) = \frac{\omega_0}{c} \cdot (n + \delta n) = \widehat{\beta}_0 + \delta\beta$$

$$\delta\beta = \frac{\omega_0}{c} \cdot \delta n = G \cdot |\tilde{E}|^2 = G'I = G''\mathcal{P} \tag{9.47}$$

$$G = k_0 n_2, \qquad G' = k_0 n_2', \qquad G'' = k_0 n_2''$$

Then, Eq. (9.13) transforms into

$$\beta(\omega) = \widehat{\beta}_0 + \delta\beta + \beta_0'(\omega - \omega_0) + \frac{1}{2} \cdot \beta_0'' \cdot (\omega - \omega_0)^2 \tag{9.48}$$

We shall replace β_0 in Section 9.1.3 by $\widehat{\beta}_0$, so we use $\beta - \widehat{\beta}_0$ instead of $\beta - \beta_0$ in Eq. (9.28). This approximately yields then, instead of Eq. (9.29), (with almost constant $\delta\beta$) the new differential equation:

$$j\frac{\partial}{\partial z}\bar{E}(z,t) = -j\beta_0' \cdot \frac{\partial}{\partial t}\bar{E}(z,t) - \frac{\beta_0''}{2} \cdot \frac{\partial^2}{\partial t^2}\bar{E}(z,t) + \delta\beta(|\bar{E}|^2) \cdot \bar{E} \tag{9.49}$$

With the transformation Eq. (9.30), we obtain

$$j \frac{\partial}{\partial \zeta} \bar{E}(\zeta, \tau) = -\frac{\beta_0''}{2} \cdot \frac{\partial^2}{\partial \tau^2} \bar{E}(\zeta, \tau) + G \cdot |\bar{E}|^2 \cdot \bar{E} \qquad (9.50)$$

The first term on the right-hand side describes the effect of dispersion, the second term the nonlinearity. Dependent on the initial pulse width σ_{t_0} and the peak power $\mathcal{P}_0$ of the incoming pulse, either the first or second effect determines the propagation of the pulse.

In order to classify the different cases, we introduce dimensionless quantities, a normalized time τ' and a normalized amplitude U:

$$\tau' = \frac{\tau}{\sigma_{t_0}} = \frac{(t - z/v_g)}{\sigma_{t_0}} \qquad (9.51)$$

$$\bar{E}(\zeta, \tau') = E_0 \cdot U(\zeta, \tau'), \qquad |U(0,0)| = 1, \qquad E_0 = |\bar{E}(0,0)|$$

Then the differential equation (9.50) transforms into

$$j \cdot \frac{\partial U}{\partial \zeta} = -\frac{\text{sign}(\beta_0'')}{2L_D} \cdot \frac{\partial^2 U}{\partial \tau'^2} + \frac{1}{L_N} \cdot |U|^2 \cdot U \qquad (9.52)$$

having the form of a nonlinear Schrödinger equation (NLS), with the two characteristic lengths

$$L_D = \frac{\sigma_{t_0}^2}{|\beta_0''|}, \qquad L_N = \frac{1}{G|E_0|^2} = \frac{1}{G'I_0} = \frac{1}{G''\mathcal{P}_0} \qquad (9.53)$$

Dependent on the relative magnitudes of the *dispersion length* $L_D = z_D/2$ [Eq. (9.39)] the *nonlinear length* L_N, and the length L of the medium, we can distinguish four cases:

1. $L \ll \min(L_N, L_D)$. Then neither the dispersive nor the nonlinear effect plays a role on a distance of length L. In Eq. (9.52), both terms on the right-hand side can be neglected, and we get $U(\zeta, \tau') = U(0, \tau')$, that is, the pulse has the same shape at $z = 0$ and $z > 0$. This case was discussed in the first paragraph of Section 9.1.4.

2. $L_D \leq L \ll L_N$. Then the last term in Eq. (9.52) is negligible. Dispersion determines the propagation of pulses; nonlinearity does not play any role. We, therefore, must have

$$\frac{L_D}{L_N} = \frac{\sigma_{t_0}^2 \cdot G''\mathcal{P}_0}{|\beta_0''|} \ll 1 \qquad (9.54)$$

This case was treated up to now in Section 9.1.4.

3. $L_N \leq L \ll L_D$. Then dispersion is negligible, in general, compared with nonlinear effects. One has

$$\frac{L_D}{L_N} = \frac{\sigma_{t0}^2 \cdot G'' \mathcal{P}_0}{|\beta_0''|} \gg 1 \tag{9.55}$$

4. $L \geq \max(L_N, L_D)$. Then both group velocity dispersion and nonlinearity play a role simultaneously.

Example 9.2

Case 1   For optical communication, we choose $L \sim 50$ km, that is, L_N, $L_D \approx 500$ km. This is guaranteed if, at $\lambda = 1.55 \ \mu$m, we choose $\beta_0'' = -20 \ \text{ps}^2/$km and $G'' = 20/(\text{km} \cdot \text{W})$ in a glass fiber, corresponding to a cross-sectional area of $A = 10 \ \mu\text{m}^2$. Then we must have $\sigma_{t_0} \geq 100$ ps and $\mathcal{P}_0 \leq 0.1$ mW.

Case 2   With the previous values of G'' and $|\beta_0''|$ for pulses of length $\sigma_{t_0} = 1$ps $\mathcal{P}_0 \ll 1$ W results . The dispersion length is $L_D = 50$ m, and with $L_N = 500$ km, we get $\mathcal{P}_0 = 0.1$ mW $\ll 1$ W.

Case 3   This can be reached easily with relatively broad pulses of $\sigma_{t_0} = 100$ ps at a peak power of $\mathcal{P}_0 \gg 0.1$ mW. The dispersion length is now $L_D = 500$ km, and with $L_N = 50$ m we find $\mathcal{P}_0 = 1$ W (for $L_N = 5$ km we would get $\mathcal{P}_0 = 10$ mW $\gg 0.1$ mW).   ∎

9.2.2   Pure Self-Phase Modulation without Dispersion: Case 3

In the case $L_N \leq L \ll L_D$, Eq. (9.52) is reduced to

$$j\frac{\partial U(\zeta, \tau')}{\partial \zeta} = \frac{1}{L_N} \cdot |U(\zeta, \tau')|^2 \cdot U(\zeta, \tau') \tag{9.56}$$

A solution of this nonlinear differential equation results from the ansatz

$$U(\zeta, \tau') = U(0, \tau') \cdot \exp[j \cdot \phi_N(\zeta, \tau')]$$

$$= U(0, \tau') \cdot \exp\left[-j\frac{\zeta}{L_N} \cdot |U(0, \tau')|^2\right] \tag{9.57}$$

$$\bar{E}(z, t) = \bar{E}(0, t) \cdot \exp\left[-j\frac{z}{L_N} \cdot \frac{|\bar{E}(0, t)|^2}{|E_0|^2}\right]$$

The pulse shape in the time domain and the intensity remain thus unchanged during propagation in the z-direction, whereas the nonlinear phase shift $\phi_N(\zeta, \tau')$ depends on the intensity and is linear in ζ:

$$\phi_N(z, t) = -\frac{z}{L_N}\left|\frac{\bar{E}(0, t)}{E_0}\right|^2 = -\frac{z}{L_N}\left|\frac{\tilde{E}(0, t)}{E_0}\right|^2 = -\frac{z}{L_N}\frac{I(0, t)}{I_0}$$

$$= -G'I(0, t) \cdot z = -\frac{\omega_0}{c}n_2'I(0, t) \cdot z = -k_0\delta n \cdot z \qquad (9.58)$$

See Figs. 9.3a and b and Eq. (9.2) in particular for a Gaussian with

$$\phi_N(z, t) = -\frac{z}{L_N}\frac{I(0, t)}{I_0} = -\frac{z}{L_N}\exp(-2a_0t^2)$$

The minimum with respect to time lies then for a Gaussian at $t = 0$, and one has

$$\phi_{N,\min} = -\frac{z}{L_N} = -\mathcal{P}_0G''z \qquad (9.59)$$

The nonlinear length L_N can thus also be characterized as that length z for which $\phi_{N,\min} = -1$.

We shall now consider an input pulse without a chirp, $b_0 = 0$. The total phase is

$$\phi_{\text{tot}} = \omega_0 t - k_0 n_{\text{tot}}z = \omega_0 t - k_0(n + \delta n)z$$
$$= \omega_0 t - k_0 nz + \phi_N(z, t) = \omega_0 t - k_0 z_{\text{tot}}^{\text{opt}}$$

with the total optical path length

$$z_{\text{tot}}^{\text{opt}} = z^{\text{opt}} + z_N^{\text{opt}} = nz + \delta n \cdot z \qquad (9.60)$$

This causes a time-dependent instantaneous frequency at the fixed position $z > 0$; see Fig. 9.3c:

$$w(z, t) = \frac{d}{dt}\phi_{\text{tot}} = \omega_0 + \frac{d}{dt}\phi_N = \omega_0 - zG'\frac{d}{dt}I(0, t)$$

$$= \omega_0 + \frac{z}{L_N}4a_0te^{-2a_0t^2} = \omega_0 + 4a_0tzG'I(0, t)$$

$$\approx \omega_0 + \frac{z}{L_N}4a_0t(1 - 2a_0t^2) \approx \omega_0 + 2b(z)t + \dots \qquad (9.61)$$

(the first row of the equation holds quite generally, the second and the third row in particular for Gaussian pulses, the approximations if a power series expansion is made). At $z = 0$, we have $w(z = 0, t) = \omega_0$; there is no chirp, $b_0 = 0$. However,

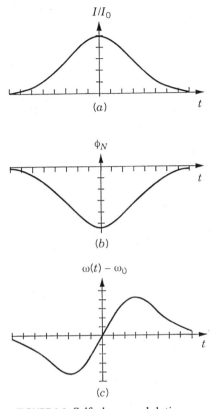

FIGURE 9.3 Self-phase modulation.

there is a chirp parameter for $z > 0$ due to nonlinearity:

$$b(z) = \frac{2a_0 z}{L_N} = 2a_0 G' I_0 z = 2a_0 z \frac{2\pi}{\lambda} n_2' I_0 > 0 \qquad (9.62)$$

since n_2' usually is positive. This is the chirp induced by the so-called *self-phase modulation*. The chirp grows in magnitude proportional to z, that is, there are always new frequency components added when propagating in the nonlinear medium. In contrast to $b(z)$, the parameter $a(z) = a_0$ has not been changed as one can see from $\bar{E}(z, t)$, Eq. (9.57).

An estimation for the spectral broadening is found, if one considers the maximum of the frequency deviation $\Delta\omega = \omega - \omega_0$ (in the case of the exact expression for the Gaussian pulses):

$$\frac{d}{dt}\Delta\omega = \frac{z}{L_N} 4a_0 e^{-2a_0 t^2}(1 - 4a_0 t^2) = 0 \qquad (9.63)$$

for $t = t_M$, that is, for

$$t_M = \frac{1}{\sqrt{4a_0}} = \sigma_{t_0}$$

or

$$\Delta\omega_{\max} = \frac{z}{L_N}\sqrt{\frac{4a_0}{e}} = -\phi_{N,\min}\frac{2}{\sqrt{e}}\sigma_\omega = |\phi_{N,\min}|\frac{2}{\sqrt{e}}\sigma_\omega \qquad (9.64)$$

where $\sigma_\omega = \sqrt{a_0}$ is the initial bandwidth according to Eq. (9.10) in the case $b_0 = 0$. The bandwidth is therefore dependent on $|\phi_{N,\min}| = \mathcal{P}_0 G''z$, that is, dependent on the input power $\mathcal{P}_0$, on the nonlinearity n_2, and on the distance z.

For $G' > 0 (n_2' > 0)$, we find according to Eq. (9.61) that for increasing z values, the frequency decreases (red shift of the visible part of the spectrum) at the leading edge of the pulse $(t < 0)$, and increases (blue shift) at the trailing edge of the pulse $(t > 0)$. Physically, this means that the contribution from the nonlinearity to the optical path length, [see Eq. (9.60)] is $z_N^{\mathrm{opt}} = \delta n \cdot z = n_2' I(0, t)z$. This contribution varies during the length of time δt:

$$\left(\frac{d}{dt}z_N^{\mathrm{opt}}\right) \cdot \delta t = \left(\frac{d}{dt}\delta n \cdot z\right) \cdot \delta t = -\frac{1}{k_0}\left(\frac{d}{dt}\phi_N\right) \cdot \delta t = -4a_0 tz n_2' I(0, t) \cdot \delta t$$

The medium appears optically longer (stretched) for $t < 0$, that is, the period will be enlarged, the frequency reduced; vice versa for $t > 0$.

The spectrum $I(z, \omega) \sim |\tilde{E}(z, \omega)|^2$ is not only broadened by self-phase modulation, but also shows an oscillating structure, where the maxima at the edges in general are also the most intensive ones. This can be explained as follows: In the $\omega(t)$ curve, there are always two values for t, for which the frequency deviations coincide exactly. This corresponds to two waves with equal frequency, but different phase, which can interfere constructively or destructively. This results in oscillations in the spectrum; see Fig. 9.4a.

The shape of the spectrum depends also considerably on the existence of an initial chirp. For a positive chirp, the number of maxima increases; for a negative one, the number decreases. The reason is this: In the central part of the pulse (near $t = 0$), the chirp is approximately linear in t with a positive slope (for $n_2 > 0$, see Fig. 9.3c). An initial chirp, [see Eq. (9.6)], must be added for $b_0 > 0$; one obtains a reinforced oscillating structure; for $b_0 < 0$, both contributions have opposite signs and cancel. For the edges of the pulse, this is not fulfilled; here, we see the remnants of the chirp (spectrum with ears). See Fig. 9.4b and c.

If two waves with different frequencies propagate in a medium, then the refractive index seen by one wave can also be changed by the intensity of the other wave. This effect is denoted as *cross-phase modulation*.

FIGURE 9.4 Oscillations in the spectrum for self-phase modulation. (Reproduced with permission from G. P. Agrawal, *Nonlinear Fiber Optics*, Academic Press, Orlando, IL., 1984.

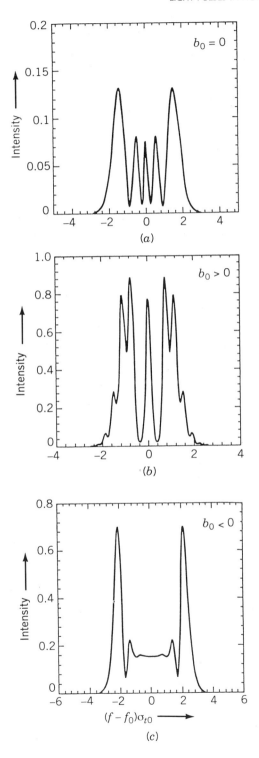

(a)

(b)

(c)

9.2.3   Self-Phase Modulation and Dispersion: Case 4

We have seen that for pure self-phase modulation, a red shift occurs at the leading edge of the pulse ($t < 0$) and a blue shift at the trailing edge of the pulse ($t > 0$). In the regime of normal dispersion, the group velocity decreases for increasing frequency; see Eq. (9.14). That is, here the red shift at $t < 0$ causes faster propagation of the leading edge and the blue shift at $t > 0$ slower propagation of the trailing edge. This means a broadening of the pulse in time. This affects again the spectral width: The previously mentioned oscillating structure can (partially) be suppressed. In the regime of anomalous dispersion ($\beta_0'' < 0$), however, v_g increases with increasing frequency. In contrast to the previous case, the temporal pulse width becomes smaller. On the other hand, in the case without nonlinearity (only dispersion effect) for large distances z, a pulse broadening is observed for both $\beta_0'' > 0$ and $\beta_0'' < 0$. See the discussion in Section 9.1.4.

The balancing between dispersion and nonlinearity can be understood as follows. In the beginning, at $z = 0$, no chirp occurs: $b_0 = 0$. Then, the chirp parameter $b(z)$ for a pure self-phase modulation is positive according to Eq. (9.62), whereas the chirp caused by dispersion, [see Eq. (9.36)], is negative for $\beta_0'' < 0$. These two chirp contributions essentially cancel; the pulse shape is adjusted automatically in such a manner that the pulse propagates practically without chirp.

Under the cooperation of self-phase modulation and dispersion, solitons can be generated: bright solitons in the regime of anomalous dispersion and dark solitons in the regime of normal dispersion. Solitons will be treated in Chapter 10.

9.2.4   Modulation Instability

In a nonlinear medium with anomalous dispersion ($\beta_0'' < 0$), a wave with constant amplitude can split up into a train of ultrashort pulses due to the simultaneous functioning of nonlinearity and dispersion.

We start with the nonlinear Schrödinger equation (9.50). At first, we look for a stationary solution with constant amplitude and τ-independent phase:

$$\bar{E} = E_0 \cdot \exp[j\phi_N(\zeta)] \tag{9.65}$$

For ϕ_N, we find after substituting Eq. (9.65) in Eq. (9.50)

$$\phi_N(\zeta) = -\zeta G E_0^2 = -\zeta G'' \mathcal{P}_0 \tag{9.66}$$

Therefore, there is a solution with constant amplitude for which the phase only depends on the power.

In order to investigate the stability of this solution, we use now the following ansatz:

$$\bar{E} = [E_0 + m(\zeta,\tau)] \cdot \exp[j\phi_N(\zeta)] \qquad (9.67)$$

with a small perturbation m. After inserting into Eq. (9.50) and neglecting quadratic terms in m, one obtains the differential equation

$$j \cdot \frac{\partial m}{\partial \zeta} = -\frac{\beta_0''}{2} \cdot \frac{\partial^2 m}{\partial \tau^2} + GE_0^2 \cdot (m + m^*) \qquad (9.68)$$

As a solution, we try, with four new constants a_1, a_2, K, Ω,

$$m(\zeta,\tau) = a_1 \cos(K\zeta - \Omega\tau) + j \cdot a_2 \cdot \sin(K\zeta - \Omega\tau) \qquad (9.69)$$

This yields two homogeneous equations for a_1 and a_2:

$$K a_1 + \frac{\beta_0''}{2}\Omega^2 a_2 = 0$$

$$\left(\frac{\beta_0''}{2}\Omega^2 + 2GE_0^2\right) a_1 + \quad K a_2 = 0$$

which possess a nontrivial solution only if for the wave number K and the angular frequency Ω of the perturbation, the following dispersion relation is satisfied:

$$K = \pm\frac{|\beta_0''|}{2}\Omega \cdot \sqrt{\Omega^2 + \text{sign}(\beta_0'') \cdot \Omega_c^2}, \qquad \Omega_c^2 := \frac{4GE_0^2}{|\beta_0''|} = \frac{4}{|\beta_0''|L_N} \qquad (9.70)$$

In the case of normal dispersion ($\beta_0'' > 0$), one has real K for all values of Ω, and the stationary solution remains stable against small perturbations. However, for anomalous dispersion ($\beta_0'' < 0$), K becomes imaginary for $|\Omega| < \Omega_c$, and the solution grows exponential, so the stationary solution, Eq. (9.65), is unstable. This spontaneous modulation of the stationary state is called *modulation instability*.

Since the wave number K of the perturbation is now imaginary, there is a gain in the power at the shifted frequency $\omega_4 = \omega + \Omega$ given by

$$g(\Omega) = 2 \cdot \text{Im}\ (K) = |\beta_0''| \cdot \Omega \cdot \sqrt{\Omega_c^2 - \Omega^2} \qquad (9.71)$$

g is real for $|\Omega| < \Omega_c$. The gain spectrum $g(\Omega)$ is skew symmetric about $\Omega = 0$ and vanishes at $\Omega = 0$ and $\Omega = \pm\Omega_c$. The extrema lie at

$$\Omega_{\max} = \pm\frac{\Omega_c}{\sqrt{2}} = \pm\sqrt{\frac{2GE_0^2}{|\beta_0''|}} = \pm\sqrt{\frac{2G''\mathcal{P}_0}{|\beta_0''|}} = \pm\frac{1}{\sigma_{t_0}}\sqrt{\frac{2\,L_D}{L_N}} \qquad (9.72)$$

with a maximum value (absolute magnitude) of

$$g_{max} = |\beta_0''| \cdot \frac{\Omega_c^2}{2} = 2GE_0^2 = 2G''\mathcal{P}_0 \tag{9.73}$$

This analysis yields only the initially exponential growth of the weak perturbation. At some time, the quadratic terms in m can no longer be neglected.

The modulation instability can be interpreted as an FWM process that is phase-matched by self-phase modulation; see also the fourth case in Section 8.1. A probe wave at $\omega_4 = \omega + \Omega$ propagates together with the stationary pump wave at $\omega_p \equiv \omega$ in the z-direction and experiences a power gain, Eq. (9.71). Two photons of the intense pump wave at ω_p will be transformed into two different photons, at ω_4 and the idler frequency $\omega_3 = 2\omega_p - \omega_4$:

$$2\omega_p = \omega_4 + (2\omega_p - \omega_4) = (\omega_p + \Omega) + (\omega_p - \Omega) \tag{9.74}$$

This simultaneous launching of pump and probe wave is called also in the literature "induced modulation instability."

PROBLEMS

Section 9.1

9.1 Evaluate $\Delta t_0, \Delta\omega$, and $\sigma_{t_0}, \sigma_\omega$ (if possible) for the normalized intensities given the following field distributions:

(a) A square pulse of width T.

(b) A double-sided exponential pulse $\exp(-|t|/T)$.

(c) A single-sided exponential pulse $\exp(-t/T)$ for $t > 0$ and $= 0$ else.

9.2 Show that a dispersive medium is a medium with memory.

9.3 Express the pulse width Δt for $b_0 = 0$ with z_D, and for $|b_0| \gg a_0 > 0$ with z_D and z_{opt}.

9.4 Derive Eq. (9.40).

9.5 Derive Eq. (9.45).

Section 9.2

9.6 Derive the change Φ_N of the phase shift due to pure self-phase modulation simply by considering the optical Kerr effect.

9.7 Calculate the optical power required for a nonlinear phase shift $\Phi_N = -\pi$ if one uses a doped fiber of length $L = 1$ m, cross section $A = 10^{-2}$ mm^2, and $n_2' = 10^{-10}$ cm^2/W at $\lambda = 1$ μm.

9.8 Derive the nonlinear phase contribution Φ_N and the frequency shift $d\Phi_N/dt$ if the Gaussian pulse at $z = 0$ is replaced by a super-Gaussian pulse:

$$\tilde{E}(0, t) = E_0 \exp j\omega_0 t \exp\left[-\frac{1+jC}{2}\left(\frac{t}{t_0}\right)^{2m}\right]$$

BIBLIOGRAPHY

G. P. Agrawal. *Nonlinear Fiber Optics*. Wiley, New York, 1984.

F. T. Arecchi and E. O. Schulz-Dubois (eds). *Laser Handbook*. Vols. 1 and 2. North-Holland, Amsterdam, 1972.

R. W. Boyd. *Nonlinear Optics*. Academic Press, San Diego, CA, 1992.

A. Hasegawa. *Optical Solitons in Fibers*. Springer-Verlag: Berlin, 1989.

A. E. Siegman. *Lasers*. University Science Books, Mill Valley, CA. 1986.

CHAPTER TEN

Solitons

We discuss here case 4 from Section 9.2.1; both dispersion and nonlinearity must be considered. Then, without restricting generality, we may even set dispersion length $L_D \approx$ nonlinear length L_N. The treatment of fundamental solitons and dark solitons can be accomplished with simple mathematics and elliptic functions. However, for N-solitons and gray solitons, the inverse scattering theory of Appendix E should be used. As supplements, we discuss dark solitons, soliton self-frequency shift, soliton-splitting, and the soliton laser.

10.1 FUNDAMENTAL SOLITON AND N-SOLITONS

As a starting point, we choose here the nonlinear Schrödinger equation (NLS) in the form of Eq. (9.52) with the variables

$$\zeta = z, \quad \tau' = \frac{\tau}{\sigma_{t_0}} = \frac{(t - z/v_g)}{\sigma_{t_o}}$$

according to Eqs. (9.30) and (9.51). At first, we consider the case of anomalous dispersion: $\beta_0'' < 0$. Therefore, we are in glass at wavelengths above the zero of dispersion $\lambda > \lambda_D \approx 1.3$ μm. Because of $n_2 > 0$, the nonlinearity constant is $G = \omega_0 n_2/c > 0$. If we set

$$\zeta' = \frac{\zeta}{(2L_D)} \approx \frac{\zeta}{(2L_N)}$$

with L_D from Eq. (9.53), we find

$$\left(j \cdot \frac{\partial}{\partial \zeta'} - \frac{\partial^2}{\partial \tau'^2} - 2|U|^2 \right) \cdot U(\tau', \zeta') = 0 \qquad (10.1)$$

This coincides with Eq. (E.24) $[\zeta' \leftrightarrow t, \tau' \leftrightarrow x]$. This PDE usually is solved with *inverse scattering theory* (IST). There, an appropriate scattering problem is considered, the potential of which is just the searched for solution. In a direct scattering problem, one knows the potential and can calculate the scattering. In the IST, one knows the scattering data (complex poles, reflection coefficients) and arrives at the "potential" $\bar{E}(\zeta, \tau')$. The IST is treated in Appendix E.

In general, one denotes a solution of the nonlinear differential equation as a solitary wave if it propagates with a constant shape or rebuilds its shape periodically in space. The word *soliton* underscores the significance of the "particle character"; it is dynamically and structurally stable, that is, if two solitons interact, they penetrate one another and travel afterward unperturbed without a change of shape. Another interpretation is that since solitons are indistinguishable, they are reflected from one another.

The case of a fundamental soliton (or single soliton) can be treated simply by the separation of variables

$$U(\tau', \zeta') = \alpha\, y(\tau')\, \exp[jg(\zeta')], \qquad \alpha > 0,\ y > 0 \tag{10.2}$$

Substituting into the differential equation (10.1) yields

$$\frac{dg}{d\zeta'} = -C = -2\alpha^2, \qquad g = -2\alpha^2 \zeta'$$

$$\frac{d^2 y}{d\tau'^2} = 2\alpha^2 y - 2\alpha^2 y^3 \tag{10.3}$$

where we have chosen as the separation constant $C = 2\alpha^2$.

If we set

$$x = \alpha\sqrt{2} \cdot \tau' \tag{10.4}$$

a normalized form of the ODE for y is obtained:

$$\frac{d^2 y}{dx^2} - y + y^3 = 0 \tag{10.5}$$

After multiplication with $dy/dx \equiv y'$, a first integration can be performed:

$$\frac{1}{2} \cdot y'^2 + \frac{y^2}{2}\left(\frac{y^2}{2} - 1\right) = W_0 \tag{10.6}$$

with an integration constant W_0. From this, we get

$$\pm \int_0^y \frac{dy_1}{\sqrt{2W_0 + y_1^2 - y_1^4/2}} = x(y) - x(0) \tag{10.7}$$

The integral $x = x(y)$ has the form of an elliptic integral. The inverse function $y = y(x)$ and, therefore, the solution of the differential equation (10.5) is then one of the 12 Jacobian elliptic functions cn (*cosinus amplitudinis*), sn, tn, dn. ... Therefore, the differential equation (10.5) is sometimes also called *cnoidal*. The elliptic functions are doubly periodic: They have a real period (as the trigonometric functions) and an imaginary one (as the hyperbolic functions). They depend on a parameter k for which one takes usually the interval $0 \leq k \leq 1$.

Instead of going into the theory of elliptic integrals and elliptic functions, we use the ansatz

$$y = A \cdot \mathrm{dn}\ (Bx|k) \tag{10.8}$$

with two constants A and B. Figure 10.1 shows the functions $\mathrm{dn}(u|k = \sqrt{0.1}\,)$ and $\mathrm{cn}(u|k = \sqrt{0.1}\,)$. Then [see (Byrd and Friedman, 1954)].

$$y' = -ABk^2 \cdot \mathrm{sn}\ (Bx|k) \cdot \mathrm{cn}\ (Bx|k)$$
$$y'' = -AB^2k^2[\mathrm{cn}^2(Bx|k) - \mathrm{sn}^2(Bx|k)] \cdot \mathrm{dn}\ (Bx|k)$$

Substituting into Eq. (10.5) and omitting the arguments Bx and k give

$$A \cdot \mathrm{dn} \cdot [-B^2k^2(\mathrm{cn}^2 - \mathrm{sn}^2) - 1 + A^2\mathrm{dn}^{\ 2}] = 0$$

Using the relations

$$\mathrm{sn}^2 + \mathrm{cn}^2 = 1, \qquad \mathrm{dn}^2 = 1 - k^2\mathrm{sn}^2$$

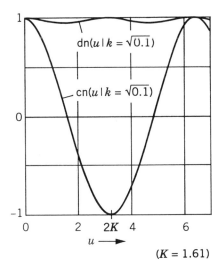

$(K = 1.61)$

FIGURE 10.1   Elliptic functions.

one finds

$$\text{sn}^2(Bx|k) \cdot (2 \cdot B^2 k^2 - A^2 k^2) + (-B^2 k^2 - 1 + A^2) = 0$$

Both coefficients must be zero; thus,

$$A = \pm\sqrt{\frac{2}{2-k^2}}, \qquad B = \pm\sqrt{\frac{1}{2-k^2}} \qquad (10.9)$$

The solution of the differential equation (10.5) is then

$$y = \pm\sqrt{\frac{2}{2-k^2}} \cdot \text{dn}\left(\frac{x}{\sqrt{2-k^2}}\bigg|k\right) \qquad (10.10)$$

The ODE (10.5) shows that with $y(x)$ also the shifted function $y(x - x_0)$ is a solution so we choose instead of Eq. (10.10) as the solution

$$y = \pm\sqrt{\frac{2}{2-k^2}} \cdot \text{dn}\left(\frac{x - x_0}{\sqrt{2-k^2}}\bigg|k\right) \qquad (10.11)$$

In the original variables of Eq. (10.2), the solution is now

$$U(\tau', \zeta') = \pm\alpha\sqrt{\frac{2}{2-k^2}} \cdot \text{dn}\left[\frac{\alpha\sqrt{2}(\tau' - \tau_0')}{\sqrt{2-k^2}}\bigg|k\right] \cdot \exp(-2j\alpha^2\zeta') \qquad (10.12)$$

Substituting Eq. (10.11) into Eq. (10.6) and using the relations

$$k^2\,\text{sn}^2 = 1 - \text{dn}^2, \qquad k^2\text{cn}^2 = \text{dn}^2 - 1 + k^2$$

we find the connection between the parameter k and the integration constant W_0:

$$W_0(k) = \frac{k^2 - 1}{(k^2 - 2)^2} \qquad (10.13)$$

or vice versa as the expression for the parameter k:

$$k^2 = \frac{4W_0 + 1 \pm \sqrt{4W_0 + 1}}{2W_0} \qquad (10.14)$$

The minimum of the left-hand side of Eq. (10.6) is obtained for $y' = 0$, $y = \pm 1$ with a value of $-1/4$. Thus $W_0 \geq -1/4$ and we can choose $0 \leq k^2 < 2$.

With Eq. (10.6), the movement of a particle with mass 1 in a force field with a

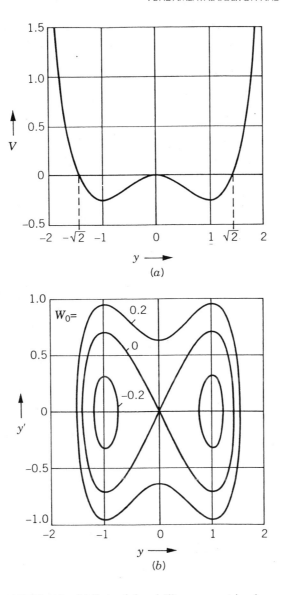

FIGURE 10.2   (*a*) Potential and (*b*) movement in phase space.

potential energy

$$V(y) = \frac{y^2}{2}\left(\frac{y^2}{2} - 1\right)$$ (10.15)

can be described. Figure 10.2*a* shows $V = V(y)$. The "total energy" W_0 is the sum of kinetic energy $y'^2/2$ and potential energy $V(y)$.

For $k = 0$, the total energy is minimal: $W_0 = -1/4$ according to Eq. (10.13). And the solution is $y = \pm 1 = \text{const}$, since $\text{dn}(u|k = 0) = 1$.

For $0 < k^2 < 1$, the total energy W_0 is still negative. Therefore, $V(y)$ must be negative too (since the kinetic energy $y'^2/2$ is always ≥ 0), that is, $0 < |y| < \sqrt{2}$. The particle oscillates in one of the two valleys.

For $k^2 = 1$, one has $W_0 = 0$ and the solution Eq. (10.11) is reduced to

$$y(x) = \sqrt{2} \cdot \text{sech}(x - x_0) \tag{10.16}$$

since $\text{dn}(u|1) = \text{sech } u = 1/\cosh u$. If the particle starts for $x - x_0 = 0$ (at "time" $\tau' - \tau'_0 = 0$) at $y = \sqrt{2}$ (or at $-\sqrt{2}$), then it takes an infinite time ($x = \infty$), in order to traverse the valley and reach $y = 0$.

For $2 > k^2 > 1 (\sqrt{2} > k > 1)$, we have $W_0 > 0$. In order that the parameter of the elliptic function does not exceed 1, one can use the relation $\text{dn}(u|k) = \text{cn}(ku|1/k)$ for $k > 1$. Thus,

$$y = \sqrt{\frac{2}{2 - k^2}} \cdot \text{cn}\left[\sqrt{\frac{k^2}{2 - k^2}} \cdot (x - x_0) \,\middle|\, \frac{1}{k}\right] \tag{10.17}$$

Now, y oscillates in finite time through the origin.

Finally, W_0 becomes arbitrarily large as $k^2 \to 2$ and the particle oscillates between $y = \infty$ and $y = -\infty$.

In phase space y', y, the relation Eq. (10.6) can be represented with W_0 as the parameter. This is shown in Fig. 10.2b for $W_0 = -0.2, 0, 0.2$.

If $k \neq 1$, there is always a periodic movement (cn or dn). A single isolated hump can be obtained only for $k = 1$ or $W_0 = 0$ with the sech function, Eq. (10.16).

This yields the stationary solution of Eq. (10.1), which is bounded and vanishes as $|\tau'| \to \infty$. Eventually, with $\tau'_0 = 0$, we find for this envelope soliton

$$U(\tau', \zeta') = -2\eta_1 \, \text{sech}(2\eta_1 \tau') \cdot e^{-4j\eta_1^2 \zeta'} \tag{10.18}$$

where we have replaced $\alpha\sqrt{2}$ by $2\eta_1$. The usual normalization is found with $2\eta_1 = -1$. Then Eq. (10.18) coincides exactly with the IST solution Eq. (E.91). $|U|^2$ that depends only on τ' is shown in Fig. 10.3a. With a translation and the Galilean transformation of Eq. (E.92), we find Eq. (E.94) that in our present notation reads

$$U(\tau', \zeta') = \text{sech}[\tau' - (\tau'_0 - 4\xi_1\zeta')] \cdot \exp j[(4\xi_1^2 - 1)\zeta' - 2\xi_1\tau'] \tag{10.19}$$

This envelope represents a wave traveling with the constant envelope velocity

$$v_E = \left(\frac{dz}{dt}\right)_E = v_g \cdot \left(1 - \frac{2\xi_1 \sigma_{t_0} v_g}{L_D}\right)^{-1}$$

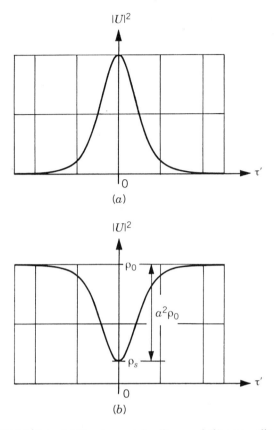

FIGURE 10.3   (*a*) Fundamental soliton and (*b*) gray soliton.

without any deformation of the profile. Equation (10.18) is the basic soliton solution: a permanent, localized wave. We have here a *fundamental* or $N = 1$-soliton, a so-called *bright soliton*. Self-phase modulation and group velocity dispersion balance in such a way that the pulse shape and pulse spectrum remain unchanged.

The N-solitons (see Fig. E.2 for $N = 2$), are solutions that (given a smooth initial shape at $z = 0$: a single hump in $-\infty < T < \infty$, and vanishing rapidly as $T \to \pm\infty$) separate for $z > 0$ into several single solitons that propagate with constant but, in general, different velocities, plus an oscillatory wave, the amplitude of which decays as the wave propagates in space. In case the initial pulse is symmetric (as, e.g., with a sech function), all the N-solitons propagate with the same velocity. The superimposed pulse shape oscillates now due to the phase interference among the solitons. The shape of the excitation periodically changes as a function of time, but all remains bounded because of the common group velocity of the individual solitons. The solitons of order N can be considered as bound states of N fundamental solitons that oscillate and form changing interference phenomena; the latter just correspond to the oscillating pulse shape.

After the so-called soliton period [see Eq. (E.95), $t_0 = \pi/4$, $t \leftrightarrow \zeta' = z/(2L_D)$] of

$$z_0 = \frac{\pi}{2} L_D = \frac{\pi}{2} \frac{\sigma_{t_0}^2}{|\beta_0''|} \tag{10.20}$$

the pulses have regained the original shape. At $\lambda = 1.55$ μm, one has $\beta_0'' = -20$ ps^2/km in glass fibers. From this results a period of $z_0 \approx 80$ m for a pulse length $\sigma_{t_0} = 1$ ps, and of $z_0 \approx 8$ km for $\sigma_{t_0} = 10$ ps. The peak power necessary for generating a soliton of order N can be calculated from $\mathcal{P}_N = N^2 \cdot \mathcal{P}_1$. The power $\mathcal{P}_1$ necessary for the fundamental soliton results from $L_D \approx L_N$ [see Eq. (9.53)], at $\lambda = 1.55$ μm (with β_0'' from above and $G'' = 20$ W^{-1} km^{-1}) and amounts to $\mathcal{P}_1 = |\beta_0''|/(G''\sigma_{t_0}^2) = 1$ W$(\sigma_{t_0}/\text{ps})^{-2}$. At $\sigma_{t_0} = 1$ ps, we get $\mathcal{P}_1 \approx 1$ W; at $\sigma_{t_0} = 10$ ps, however, $\mathcal{P}_1 = 10$ mW. For the higher powers, necessary for solitons of higher order, at first the nonlinear effect predominates; however, dispersion soon catches up. This results in the periodic oscillations with the period z_0.

10.2   SUPPLEMENTS

10.2.1   Dark Solitons

Whereas for $\lambda > \lambda_D$ the group velocity dispersion is $\beta_0'' < 0$, we have $\beta_0'' > 0$ for $\lambda < \lambda_D$, in the region of anomalous dispersion we are now interested in. In contrast to the treatment in Section 10.1, we have here sign $(\beta_0'') = 1$ and obtain instead of Eq. (10.1) as the nonlinear Schrödinger equation

$$\left(j\frac{\partial}{\partial \zeta'} + \frac{\partial}{\partial \tau'^2} - 2 \cdot |U|^2 \right) \cdot U(\tau', \zeta') = 0 \tag{10.21}$$

Of course, the solution here can also be found by inverse scattering theory. However, for simplicity, we try the following separation of variables:

$$U = \sqrt{\rho(\tau')} \cdot \exp[j\sigma(\zeta')], \qquad \rho \geq 0 \tag{10.22}$$

resulting in two differential equations (we are omitting the primes for simplicity)

$$\frac{d\sigma}{d\zeta} = \kappa, \qquad \text{or} \qquad \sigma = \kappa\zeta$$

$$0 = \frac{d^2\rho}{d\tau^2} \rho - \frac{1}{2}\left(\frac{d\rho}{d\tau}\right)^2 - 4\rho^3 - 2\kappa\rho^2 \tag{10.23}$$

where κ is a separation constant. In the second ODE, the transformation

$$\frac{d\rho}{d\tau} \equiv p = p(\rho)$$

leads to a Bernoulli differential equation

$$\frac{dp}{d\rho} - \frac{p}{2\rho} - \frac{4\rho^2}{p} - 2\kappa \frac{\rho}{p} = 0$$

and the transformation $u = p^2$ to the inhomogeneous linear ODE of first order

$$\frac{du}{d\rho} - \frac{u}{\rho} - (8\rho^2 + 4\kappa\rho) = 0$$

with the solution

$$u = p^2 = \left(\frac{d\rho}{d\tau}\right)^2 = 4\rho(C + \rho^2 + \kappa\rho) \tag{10.24}$$

where C is an integration constant. We want now a solution $\rho(\tau)$ for the dark soliton, which has a minimum $(d\rho/d\tau = 0)$ for $\rho = \rho_{min} = 0$ and asymptotically a maximum for $\rho = \rho_0$, that is, $(d\rho/d\tau)^2$ from Eq. (10.24) must have the form

$$\left(\frac{d\rho}{d\tau}\right)^2 = 4\rho(\rho - \rho_0)^2, \qquad \rho_0 = -\frac{\kappa}{2}, \qquad C = \rho_0^2 = \frac{\kappa^2}{4} \tag{10.25}$$

This ODE can be integrated easily and we obtain

$$\rho(\tau) = \rho_0 \tanh^2(\sqrt{\rho_0}\tau) = \rho_0[1 - \text{sech}^2(\sqrt{\rho_0}\tau)] \tag{10.26}$$

as expected. This represents a *dark soliton,* or a *topological soliton.*

However, solutions also exist in which the minimum value is not $\rho_{min} = 0$ but $\rho_{min} = \rho_s > 0$. Then the simple separation ansatz, Eq. (10.22), does not work since we have now the additional parameter ρ_s, and we should replace the former solution by

$$\rho(\tau) = \rho_0[1 - a^2 \text{ sech}^2(\sqrt{\rho_0}a\tau)], \qquad a^2 = 1 - \frac{\rho_s}{\rho_0} \tag{10.27}$$

representing a "gray" soliton. The total solution is then

$$U(\tau', \zeta') = \sqrt{\rho(\tau')} \cdot \exp[j\sigma(\zeta', \tau')\zeta'] \tag{10.28}$$

where the function $\sigma(\zeta', \tau')$ is now a complicated function of ζ' and τ'. This solution means a dent at $\tau' = 0$ in an asymptotically constant wave train with amplitude ρ_0. The dent has a minimum value of $\rho_s = \rho_0(1 - a^2)$. It is this "missing" light that is denoted as a dark soliton or gray soliton, respectively. Fig.10.3b shows $|U|^2$ for a gray soliton. These kinds of solitons have not found any application up to now.

10.2.2 Soliton Self-Frequency Shift and Soliton Splitting

There are two nonlinear effects of higher order than those contained in Eq. (10.1): Raman-induced soliton self-frequency shift and soliton splitting.

If the soliton pulse width lies below 1 ps, the induced Raman effect takes care of an energy redistribution inside the pulses: An arbitrary Fourier component of the soliton can be considered a pump wave, which hits a Raman-active medium (glass). The Raman amplification profile tells us by what amount the frequency parts with lower frequencies are amplified. Altogether, one thus obtains a transport of energy inside the soliton from higher to lower frequencies. Thereby, the pulse gets its midfrequency shifted to somewhat lower values. At this self-pumped frequency shift, the soliton again keeps its shape.

Due to

$$\Delta v_g = \left(\frac{\partial v_g}{\partial \omega} \right) \cdot \Delta \omega = -\left(\frac{\beta_0''}{\beta_0'^2} \right) \cdot \Delta \omega \tag{10.29}$$

[see Eq. (9.14)] and for $\beta_0'' < 0$, the mentioned negative frequency shift is coupled with a decreasing group velocity. Without the self-induced Raman effect, all the N-solitons ($N > 1$) would propagate with the same velocity. However, if one takes into account the self-induced Raman effect, they obtain different velocities and separate spatially.

10.2.3 The Soliton Laser

We use an optical resonator, in which we put, for example, a color-center crystal pumped by a mode-coupled Nd:YAG laser; see Fig. 10.4. Through one of the two resonator mirrors light comes out, which is launched into a (polarization-preserving) monomode fiber, back-reflected at the end of the fiber, and then it enters anew the resonator. In fact, the soliton laser uses $N = 2$-solitons. The length of the fiber is chosen as approximately half the soliton period. Thus, the

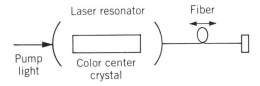

FIGURE 10.4 Soliton laser.

pulse returning from the fiber has the same width (or somewhat smaller) as the pulse that was launched into the fiber.

A pulse coming from the resonator has at first a certain width, whereas that coming back from the fiber is somewhat smaller, exciting the laser to an emission of a shorter pulse (since the laser effect in the soliton laser comes mainly from the injected pulse), and so on. A stationary operation occurs if the in- and outgoing pulses have the same length. As a matter of fact, the pulse will be broadened at each round-trip in the resonator by dispersive elements. Therefore, for compensation, the pulse must be slightly comprised in the fiber. This is not possible with a fundamental soliton, but rather with an $N = 2$-soliton. With the soliton laser, one has a stable source of Fourier-limited pulses of arbitrary duration. For example, pulses of 50-fs width at $\lambda = 1.5$ μm have been produced with a soliton laser.

PROBLEMS

See Appendix E.

REFERENCES

P. F. Byrd and M. D. Friedman. *Handbook of Elliptic Integrals for Engineers and Physicists*. Springer-Verlag, Berlin, 1954.

BIBLIOGRAPHY

G. P. Agrawal. *Nonlinear Fiber Optics*. Academic Press, Boston, MA, 1989.

J. M. Cervero. Unveiling the solitons mystery: The Jacobi elliptic functions. *Am. J. Phys.* 54 (1986): 35–38.

A. Hasegawa. *Optical Solitons in Fibers*. Springer-Verlag, Berlin, 1989.

A. C. Scott, F. Y. F. Chu and D. W. McLaughlin. The soliton: A new concept in applied science. *Proc. IEEE* 61 (1973): 1443–1485.

CHAPTER ELEVEN

Nonlinear Effects in Glass Fibers

Nonlinear effects can be disadvantageous in optical communication and represent a limit for the maximum power allowable in a glass fiber. On the other hand, the same effects can be useful for fiber-optical amplifiers, oscillators, and modulators and for the compensation of pulse dispersion (soliton). Glass has, as an amorphous material, an inversion center. Therefore, three-wave mixing is forbidden. Nevertheless, three-wave mixing can occur due to an electric quadrupole moment or a magnetic dipole moment, however weak.

11.1 THE HARMFUL EFFECTS

The *induced Raman effect* (IRE) describes the scattering of photons from "optical" phonons in solids; see Chapter 6. Glass fibers from quartz glass are an amorphous material. The vibrational frequencies of the molecules are washed out to bands that overlap and constitute a continuum. So the Raman gain in glass fibers extends continuously over a large range (this makes possible broadband amplification). The IRE leads to the generation of Stokes waves, the frequencies of which are determined by the maximum of the Raman gain. The frequency shift of the Stokes and the anti-Stokes lines is about Δf (phonon) $\approx 1-10$ THz ($\Delta \lambda \approx 0.1$ μm at $\lambda = 1.5$ μm).

For a constant pump, Eq. (6.37) yields an exponential growth of the power of the Stokes wave at frequency f_s on a distance $z = L$ according to

$$\mathcal{P}_s(L) = \mathcal{P}_s(0) \cdot \exp[g_r \cdot |\overline{E}(0, f_p)|^2 \cdot L] = \mathcal{P}_s(0) \cdot \exp\left(\frac{g_r' \cdot \mathcal{P}_p \cdot L}{A}\right) \quad (11.1)$$

where $\mathcal{P}_p$ is the pump power at frequency f_p. The Raman gain factor g_r varies linearly according to Eq. (6.33) with the frequency f_s. [We assume that χ and n are constant. The connection between g_r, g_r', g_r'' is exactly as in Eq. (9.46).] If there are losses in the glass medium, one has to supplement the coupled system of differential equations (6.34) by damping terms with the absorption coefficients α_s, α_p at the Stokes and pump frequency:

$$\frac{dI_s}{dz} = g'_r \cdot I_p I_s - \alpha_s I_s$$

$$\frac{dI_p}{dz} = -\frac{\omega_p}{\omega_s} \cdot g'_r \cdot I_p I_s - \alpha_p I_p \tag{11.2}$$

If pump depletion is neglected, the first term on the right-hand side of the second equation will be omitted and I_p varies only due to the damping in the distance z:

$$I_p(z) = I_0 \cdot \exp(-\alpha_p z)$$

If this is substituted in the first differential equation, this equation also can be solved:

$$I_s(L) = I_s(0) \cdot \exp(g'_r \cdot I_0 \cdot L_e) \cdot \exp(-\alpha_s L) \tag{11.3}$$

with

$$L_e = \frac{[1 - \exp(-\alpha_p L)]}{\alpha_p} \tag{11.4}$$

as an effective length calculable from the actual fiber length L and the linear absorption α_p. (The equations can also be solved exactly for pump depletion if one can set $\alpha_p \approx \alpha_s$.)

To avoid this nonlinear effect, the pump power $\mathcal{P}_p$ must be sufficiently small. The Raman threshold power $\mathcal{P}_c$ is defined as that pump power for which the Stokes wave at the fiber end is already as large as the pump wave. With the assumption of a gain spectrum of the form of a Lorentzian profile, [see also Eq. (6.36)], one can show (Smith, 1972) that this threshold condition is approximately fulfilled if the exponent in Eq. (11.1) has a value of 16:

$$\mathcal{P}_c = \frac{16 \cdot A}{(g'_r \cdot L_e)} \tag{11.5}$$

For IRE in the backward direction, the threshold is a little higher; analysis shows that one has to replace the factor 16 by 20.

The threshold power for a low-loss monomode fiber with $\lambda_0 = 1.5$ μm, $A = 70$ μm^2, $L = 20$ km, and $g'_r = 7 \cdot 10^{-14}$ m/W can be calculated as $\mathcal{P}_c = 800$ mW. This is a value that appears harmless for optical communication. However, in a wavelength division multiplex (WDM) system, the Raman amplification is dangerous at already much lower power since the energy can be transferred to a longer-wavelength channel. This Raman-induced crosstalk should not be larger than $1\% \cong 20$ dB. From this follows:

$$\frac{\mathcal{P}_s(L)}{\mathcal{P}_s(0)} = \exp\left(g_r' \cdot \mathcal{P}_p \cdot \frac{L}{A}\right) = 1 + 0.01, \qquad g_r' \cdot \mathcal{P}_p \cdot \frac{L}{A} \approx 0.01$$

and the value of 16 in Eq. (11.2) has to be reduced to 0.01. Thereby, the critical power is lowered to an order of magnitude of 1 mW. One should not choose the channel separation just equal to a value that corresponds to the maximal Raman gain; from this it follows that a separation of the different wavelength channels is required of more than 500 cm^{-1} (this corresponds to a $\Delta\lambda$ of more than 0.1 μm at 1.5 μm).

The *induced Brillouin effect* (IBE) describes the scattering of photons from "acoustical" phonons. Since these lie in the frequency range of $10^9 \ldots 10^{10}$ Hz, a frequency shift much smaller than with IRE occurs. Owing to momentum conservation, the IBE vanishes in the forward direction, in the backward direction, it is maximum. Nevertheless, spontaneous Brillouin scattering in the forward direction can occur: "Brillouin scattering by guided acoustic waves." Due to the guidance in the wave guide, the wave vector selection rule is weakened. Although the effect is very weak, it has played a disturbing role, for example, in the generation of nonclassical light (squeezed light) in glass fibers.

In the case of damping in the glass fiber, Eq. (6.43) is to be supplemented again as follows:

$$\frac{dI_b}{dz} = - g_b' \cdot I_p I_b + \alpha I_b$$

$$\frac{dI_p}{dz} = - g_b' \cdot I_p I_b - \alpha I_p$$
(11.6)

Since $\omega_b \approx \omega_p$, we have also $\alpha_b \approx \alpha_p =: \alpha$. The reverse sign in the first equation compared with Eq. (11.2) stems again from the backward-propagating wave I_b. For a "constant" pump $I_p = I_0$, we again neglect in the second Eq. (11.6) the first term on the right-hand side and consider only the damping: $I_p(z) = I_0 \exp(-\alpha z)$. Substituting into the first Eq. (11.6) and using again L_e from Eq. (11.4) result in

$$I_b(0) = I_b(L) \cdot \exp\left(\frac{g_b' \cdot \mathcal{P}_p \cdot L_e}{A}\right) \exp(-\alpha L)$$
(11.7)

The gain for the IBE in glass can be larger than for the IRE up to a factor of 300. Instead of Eq. (11.5), one now has to take for the critical pump power which is defined similarly as the Raman threshold power (Smith, 1972)

$$\mathcal{P}_c = \frac{21 \cdot A}{(g_b' \cdot L_e)}$$
(11.8)

For a large (compared with the pump linewidth) Brillouin linewidth, $\Delta f_b \approx 100$ MHz, one finds $g_b' = 5 \cdot 10^{-11}$ m/W at $\lambda_p = 1.55$ μm. With $A = 50$ μm^2, $L_e = 20$ km, this results in $\mathcal{P}_c \approx 1$ mW. This is the reason for the IBE being the dominant nonlinear process in glass fibers.

The harmful effects of the IBE represent therefore a considerable signal reduction; they cause multiple-frequency shifts and produce an intense backward wave.

The disturbing IBE can (partially) be remedied by different measures (in particular by raising the threshold power $\mathcal{P}_c$):

1. By using short pulses or relatively broadband-pump light sources.
2. By exposing the fiber to a large-temperature gradient (the fiber is wound on a drum; the lower half of the drum with a horizontally lying axis is cooled while the upper half is heated); this effects a broadening of the gain curve via the temperature-dependent sound velocity while having a constant area, that is, the maximum Brillouin gain is lowered and the threshold raised.
3. By an appropriate digital modulation: amplitude shift keying (ASK) and frequency shift keying (FSK) raise the critical power only a few; with phase shift keying (PSK), however, a whole order of magnitude can be gained.

Due to the minor frequency shift, IBE is not as dangerous for WDM as IRE.

A very dangerous effect for optical communication is *self-phase modulation*, which was discussed in Section 9.2.2. For high light intensities, the refractive index of glass has a Kerr nonlinearity of the form

$$n = n_0 + n_2' I \tag{11.9}$$

Equation (9.61) shows that in spite of a small value for n_2, a considerable frequency shift can occur, if only the length z of the glass fiber is sufficiently large. As critical power, we choose here that power at which the time-bandwidth product, Eq. (9.11), has grown by a factor of $\sqrt{2}$. Concerning this, one can use Eq. (9.62), which shows that the chirp parameter is increased from $b_0 = 0$ to

$$b(L) = \frac{4\pi}{\lambda} a_0 L n_2' I_0$$

while we assume that the pulse shape has not changed: $a(L) = a_0$. For $b(L) = a(L)$, the critical power $\mathcal{P}_c = I_0 A$ is reached, with A as the effective cross section of the glass fiber. We thus obtain

$$\mathcal{P}_c = \frac{A \cdot \lambda}{(4\pi \cdot L \cdot n_2')} \tag{11.10}$$

With $\lambda = 1.5$ μm, $A = 10$ μm^2, $L = 2$ km, $n_2' = 3.2 \times 10^{-20}$ m^2/W, one finds $\mathcal{P}_c = 20$ mW.

The *modulation instability* discussed in Section 9.2.4 is a harmful effect: It can be a limiting factor for coherent optical communication because of the undesired amplitude modulation. At first, we obtain from Eq. (9.70) with $\beta_0'' = -20$ ps^2/km, $G'' = 2$ W^{-1} km^{-1}, a maximal frequency shift of $\Delta f = \pm\Omega_c/(2\pi) = \pm100$ GHz for a power $\mathcal{P}_p = 1$ W. Analogously to the treatment of the IRE, we get from the maximum power gain, defined by

$$\frac{\mathcal{P}_s}{\mathcal{P}_p} = \exp(g_{\max} \cdot z), \qquad g_{\max} = 2G'' \cdot \mathcal{P}_p$$

according to Eq. (9.73), a power of 2.5 mW after a length of 1 km for an admissible crosstalk of 20 dB/km.

In *four-wave mixing*, three waves at the frequencies ω_1, ω_2, ω_3 generate a fourth one at the frequency ω_4. Which of the many different FWM processes occurs depends on the phase relations among the fields, and these are determined by the conservation of energy and momentum. In contrast to IRE and IBE, where phase matching is automatically fulfilled as a result of the active participation of the nonlinear medium in the process (molecular vibrations or acoustical phonons), the phase-matching condition $\Delta\mathbf{k} = 0$ in parametric processes, where the material only plays a passive role, must be (approximately) explicitly fulfilled in each case.

The basic relations for the process $f_4 = 2f_1 - f_3$ have been derived already in the fourth case in Section 8.1. The refractive indices n_i must now be replaced in the wave guide by effective refractive indices

$$n_i \rightarrow \tilde{n}_i = n_i + \Delta n_i \tag{11.11}$$

and Δn_i is the change of the index of refraction due to wave guiding. Thus, finally one has to set also

$$\Delta k = \Delta k_m + \Delta k_w \tag{11.12}$$

where, according to Eq. (8.16),

$$\Delta k_m = 2\pi \frac{(f_4 n_4 + f_3 n_3 - 2f_1 n_1)}{c}$$

$$\Delta k_w = 2\pi \frac{[f_4 \Delta n_4 + f_3 \Delta n_3 - f_1(\Delta n_1 + \Delta n_2)]}{c} \tag{11.13}$$

Δk_m comes from the material dispersion and Δk_w from the wave guide dispersion. *Note*: Although the two pump waves E_1, E_2 posses the same frequency, Δn_1, Δn_2 are in general different from each other if E_1, E_2 propagate in different modes. Moreover, one has the following relation, as can be seen from an expansion of $k_i \equiv \beta(\omega_i)$ about the frequency $f_1 = f_2 = f_p$ [see Eq. (9.13)]:

$$\Delta k_m = \beta(\omega_4) + \beta(\omega_3) - 2\beta(\omega_1) = \frac{\beta_1''}{2} \cdot [(\omega_4 - \omega_1)^2 + (\omega_3 - \omega_1)^2] = \beta_1'' \Omega_s^2$$

$$(11.14)$$

Here, the Eqs. (8.16) and (8.17) have been used. As an effective propagation constant, the quantity

$$\kappa = \Delta k + \Delta k_{NL} = \Delta k_m + \Delta k_w + \Delta k_{NL} \qquad (11.15)$$

appears. The nonlinear part is proportional to $|\overline{E}_1|^2$ or $\mathcal{P}_1$:

$$\Delta k_{NL} = \delta\beta = G'' \cdot 2\mathcal{P}_1 \qquad (11.16)$$

See Eq. (9.47). The factor of 2 comes from the two pump waves $\mathcal{P}_1 + \mathcal{P}_2$ with $\mathcal{P}_1 \equiv \mathcal{P}_2$. Finally, $\Delta k_{NL} = K$ according to Eq. (8.19).

In the expression for the parametric gain, Eq. (8.21), Δk must still be replaced by κ. Then, the parametric gain is given by

$$g = \sqrt{(2G''\mathcal{P}_1)^2 - \left(\frac{\kappa}{2}\right)^2} = \frac{1}{2}\sqrt{3(\Delta k_{NL})^2 - \Delta k^2 - 2 \cdot \Delta k \cdot \Delta k_{NL}} \qquad (11.17)$$

In the range

$$-6G''P_1 \le \Delta k \le 2G''P_1$$

the radicand is positive, so a gain occurs. The maximum gain is

$$g_{max} = 2G''\mathcal{P}_1 = \Delta k_{NL}$$

It is reached for $\kappa = 0$ or $\Delta k = -\Delta k_{NL}$ (due to self-phase modulation not for $\Delta k = 0$); this means that for $\kappa = 0$, at least one of the three contributions in Eq. (11.15) must be negative. The maximal parametric gain is larger by a factor of 2 than the Raman gain in the maximum, so the threshold is also lower here than with the IRE. Nevertheless, the IRE dominates in general because it is difficult for the FWM in long fibers to maintain phase matching because of fluctuations of the core diameter. As the *coherence length* for FWM in fibers, one can define [similar to the case of frequency doubling, Eq. (4.7)]

$$L_{coh} = \frac{2\pi}{\kappa_{max}} \qquad (11.18)$$

where κ_{max} is the maximum tolerable phase mismatch.

Monomode fibers   Here, Δk_w stems from the dispersion of the single guided mode; therefore, it is very small (all Δn_i are approximately equal) and will be neglected. For λ_p below the zero of dispersion $\lambda_D \approx 1.3\,\mu m\ (\beta_1'' > 0)$, one has to keep both Δk_m and Δk_{NL} small separately by using small frequency shifts Ω_s and low pump powers. If the part Δk_m dominates,

$$L_{\text{coh}} \approx \frac{2\pi}{|\Delta k_m|} = \frac{2\pi}{\beta_1''\Omega_s^2} \tag{11.19}$$

results. In the visible range, one has $\beta_1'' \approx 50\text{--}60\ \text{ps}^2/\text{km}$, resulting in a coherence length of $L_{\text{coh}} \approx 1\ \text{km}$ for a frequency shift of $f_s = \Omega_s/2\pi \approx 50\ \text{GHz}$. This means that a significant FWM occurs in monomode fibers for such frequency shifts f_s, for which $L \leq L_{\text{coh}}$ is. This can happen in WDM communication systems. One can obtain considerable crosstalk. Thus, for a monomode fiber with 5-GHz channel spacing and -20-dB allowed crosstalk, the input power per channel should be smaller than 0 to -5 dBm (1 mW...0.316 mW) for a 15 km-long fiber, and -10 dBm (0.1 mW) for 100 km (for a low-loss fiber dispersion-shifted to $1.5\,\mu m$). However, this effect can be avoided by an appropriate frequency separation of the channels.

For a pump wavelength $\lambda_1 > \lambda_D\ (\beta_1'' < 0)$, the negative value of Δk_m must be compensated by Δk_{NL} (Δk_{NL} must be made sufficiently large). For phase matching, one has

$$\kappa = 0 = \Delta k + \Delta k_{NL} \approx \Delta k_m + \Delta k_{NL} = \beta_1''\Omega_s^2 + \Delta k_{NL}$$

From this, we get

$$\Omega_s = \sqrt{\frac{2G''\mathcal{P}_1}{|\beta_1''|}} \tag{11.20}$$

coinciding exactly with the frequency shift $\Omega_{\max}$ of Eq. (9.72) of the modulation instability. Modulation instability can be interpreted in the frequency domain as FWM, whereas in the time domain, it can be described by the unstable growth of a small perturbation.

Multimode fibers   Since $\beta_1'' > 0$ for a pump wavelength $\lambda_p < \lambda_D$, one has $\Delta k_m > 0$ in the visible range. Δk_w must be made negative by propagating the waves in different modes.

Birefringent fibers   The variable, κ, also contains the three previous terms, but Δk_{NL} is different from Eq. (11.16), and a main contribution to Δk_w is now the modal birefringence

$$\delta n = \Delta n_x - \Delta n_y \tag{11.21}$$

if the waves propagate in two orthogonally polarized modes. The variations Δn_x, Δn_y of the refractive index refer to the slow (Δn_x large) and fast (Δn_y small) fiber axis. If one neglects Δk_{NL}, then Δk_w must be negative for $\lambda_p < \lambda_D$, and this can be achieved if the pump wave is polarized along the slow axis ($\Delta n_1 = \Delta n_2 = \Delta n_x$) and the Stokes and anti-Stokes waves along the fast axis ($\Delta n_3 = \Delta n_4 = \Delta n_y$). So, with the help of the frequency relation $\omega_3 + \omega_4 = 2\omega_p$, one obtains

$$\Delta k_w = \frac{1}{c}[\Delta n_y(\omega_3 + \omega_4) - 2 \cdot \Delta n_x \cdot \omega_p] = -2\frac{\omega_p}{c}(\Delta n_x - \Delta n_y) = -2\frac{\omega_p}{c}\delta n$$

. (11.22)

From the compensation condition $\Delta k_m = -\Delta k_w$ then follows:

$$\Omega_s = \sqrt{\frac{4\pi \cdot \delta n}{\lambda_p \cdot \beta_1''}} \tag{11.23}$$

At a pump wavelength of $\lambda_p = 0.532\,\mu m$, $\beta_1'' = 60\,ps^2/km$, and a typical value of $\delta n = 10^{-5}$ for the fiber birefringence, a frequency shift of $f_s = \Omega_s/(2\pi) \sim 10\,THz$ results.

11.2   THE USEFUL EFFECTS

The *induced Raman effect* is the most frequently used effect for optical amplification mainly because of the broad gain-bandwidth. It follows that there is practically no limit for the communication speed of the signal to be amplified (a very small Δt is allowed), and that one can amplify with a single powerful pump in several wavelength channels simultaneously. The pump wave and signal to be amplified can propagate co- or contradirectionally.

A pump wave at frequency ω_p is propagating in the fiber. If there is also a signal wave at ω_s, it will be amplified due to the Raman gain g_r, as long as $\omega_p - \omega_s$ lies within the Raman gain spectrum. If only the pump is incident, the spontaneous Raman scattering yields a weak signal ω_s, which then can be amplified subsequently. It is true that all frequency components are amplified, but most of all, those components for which g_r is maximum. Therefore, $\omega_p - \omega_s$ should lie close to the maximum of the Raman gain curve. Starting from the threshold, this component will be built up nearly exponentially.

Without a pump, $I_p = 0$, one should use for the signal wave

$$I_s(L) = I_s(0) \cdot \exp(-\alpha_s L)$$

according to Eq. (11.2). Thus, the amplification factor becomes, using Eq. (11.3),

$$G_a = \frac{I_s(L) \text{ with pump}}{I_s(L) \text{ without pump}} = \exp\left(g_r' \cdot \mathcal{P}_p \cdot \frac{L_e}{A}\right) \tag{11.24}$$

With $g_r' = 10^{-13}\,m/W$, $L_e = 100\,m$, $A = 10\,\mu m^2$, one can see that for a pump power of $\mathcal{P}_p \geq 1\,W$, already considerable amplification begins: $G = \exp(1)$. But, to be sure, very soon one has to again take into account pump depletion. With a

1.33 μm Nd:YAG laser as the pump, a cw-signal amplification up to 20 dB can be achieved. Perturbing influences a rise from the spontaneous Raman scattering and low-frequency fluctuations in the pump source, which can again grow very strongly due to IBE. However, there are also various attempts to suppress IBE.

A short time ago, only Nd:YAG and color-center lasers were taken into consideration as pump lasers. Nowadays, people use also monomode fibers doped with rare earths as optical amplifiers. Erbium is very attractive, since it has strongly fluorescing peaks at 1.536 μm, which lie very close to the damping minimum in conventional glass fibers.

If a Raman-active material, which lies with its Stokes-shifted wavelength still within the gain spectrum, is put into an optical resonator with sufficient feedback at the Stokes wavelength λ_s and if another laser is used as the pump, the losses at a full round-trip can be compensated by the Raman gain. A strong optical oscillation is generated at λ_s: This is the *Raman laser*. Using a glass prism inside the resonator, one can select a desired wavelength. A tuning range of about 10 nm ($\sim$10 THz) is achieved. In particular, in a fiber Raman laser, a Raman-active fiber is used in the resonator. We have then a distributed optical amplification of the optical signal pumped by the laser light propagating in the fiber in either the same or opposite direction. Owing to the feedback, the threshold pump power is lowered considerably compared with the one-way path; see Eq. (11.5).

By exploiting the *induced Brillouin effect*, fiber Brillouin lasers (with a ring or a Fabry–Perot-geometry) have also been built, moreover fiber Brillouin amplifiers: By this means, for example, the repeater distance in optical communication can be extended, or the sensitivity of detection raised, or only a part of the spectrum selectively amplified.

Four-wave mixing is used, for example, in a fiber-optical parametric amplifier. In the case in which a signal wave launched at $z = 0$, with $\mathcal{P}_3(z = 0) \neq 0$, is amplified and simultaneously an idler wave appears, with $\mathcal{P}_4(z = 0) = 0$, the solution Eq. (8.20) leads, with $\Delta k \to \kappa$ and if we take into account Eq. (11.17), to the equations

$$\mathcal{P}_3(z) = \mathcal{P}_3(0) \cdot \left[1 + \left(1 + \frac{\kappa^2}{4g^2}\right) \cdot \sinh^2(gz)\right] \tag{11.25}$$

$$\mathcal{P}_4(z) = \mathcal{P}_3(0) \cdot \left(1 + \frac{\kappa^2}{4g^2}\right) \cdot \sinh^2(gz) = \mathcal{P}_3(z) - \mathcal{P}_3(0)$$

Then the one-way gain of the parametric amplifier becomes, with Eq. (11.17),

$$G_a = \frac{\mathcal{P}_3(L)}{\mathcal{P}_3(0)} = 1 + \frac{\mathcal{P}_4(L)}{\mathcal{P}_3(0)} = 1 + \left(1 + \frac{\kappa^2}{4g^2}\right) \cdot \sinh^2(gL) \tag{11.26}$$

$$= 1 + \frac{(2G''\mathcal{P}_1)^2}{g^2} \cdot \sinh^2(gL) \approx \left(1 + \frac{\kappa^2}{4g^2}\right) \cdot \sinh^2(gL)$$

The approximation holds for $gL \gg 1$. The amplifier gain now depends on κ. Only in the case that

$$|\kappa| < 4G''\mathcal{P}_1 = 2 \cdot \Delta k_{NL}$$

amplification $g > 0$ exists according to Eq. (11.17). This also means good phase matching, since $|\kappa|$ must be small then. If one has, in particular, complete phase matching, $\kappa = 0, g = 2G''\mathcal{P}_1$, and still $gL \gg 1$, one obtains exponential behavior for G_a:

$$G_a = \sinh^2(gL) \approx \frac{1}{4} \cdot \exp(2gL) = \frac{1}{4} \cdot \exp(4G''\mathcal{P}_1 L) \tag{11.27}$$

As a general connection between phase mismatch κ and frequency shift Ω, one has, with

$$\kappa = \beta_1'' \cdot \Omega_s^2 + \Delta k_w + \Delta k_{NL}$$

[see Eqs. (11.14) and (11.15)]

$$\frac{\Delta\kappa}{\Delta\Omega_s} \approx \left| \frac{\partial\kappa}{\partial\Omega_s} \right| = 2|\beta_1''| \cdot \Omega_s, \qquad \Delta\Omega_s = \frac{\Delta\kappa}{2 \cdot |\beta_1''| \cdot \Omega_s} \tag{11.28}$$

In order to estimate the amplification bandwidth $\Delta\Omega_s$, we set

$$\Delta\kappa = \kappa_{\max} = 4G''\mathcal{P}_1$$

in the case of amplification. From this results

$$\Delta\Omega_s = \frac{2G''\mathcal{P}_1}{|\beta_1''| \cdot \Omega_s} = \frac{2}{L_N \cdot |\beta_1''| \cdot \Omega_s} \tag{11.29}$$

Amplification requires high pump power; therefore, L_N is small ($L_N \ll L$). For $|\beta_1''| = 20 \ldots 60 \text{ ps}^2/\text{km}$, $\Omega_s = 2\pi \times (10 \ldots 100) \times 10^{12} \text{ s}^{-1}$, and $L_N = 1\,\text{m}$, one finds $\Delta f_s = \Delta\Omega_s/2\pi$ in the order of magnitude of 10–100 GHz. This bandwidth lies between that of a fiber Raman amplifier (5 THz) and a fiber Brillouin amplifier (100 MHz).

For $|\kappa| > 4G''\mathcal{P}_1$, however, g becomes imaginary:

$$g = j \cdot \tilde{g}, \qquad \tilde{g} = \sqrt{\left(\frac{\kappa}{2}\right)^2 - (2G''\mathcal{P}_1)^2} \tag{11.30}$$

There is no longer an amplification, but rather a periodical power variation at the waves ω_3 and ω_4. If in particular $|\kappa| \gg 4G''\mathcal{P}_1$, then $\tilde{g} \approx \kappa/2$ and

$$P_4(L) = P_3(0) \left[g^2 + \left(\frac{\kappa}{2} \right)^2 \right] L^2 \frac{\sinh^2 j\tilde{g}L}{(j\tilde{g}L)^2} = P_3(0)(2G''P_1L)^2 \cdot \frac{\sin^2(\tilde{g}L)}{(\tilde{g}L)^2}$$

$$= P_3(0) \cdot (2G''P_1L)^2 \cdot \frac{\sin^2(\kappa L/2)}{(\kappa L/2)^2} \tag{11.31}$$

We now set $L_w = 2\pi/\Delta\kappa$, representing a measure for the width. Thus, as mixing bandwidth from Eq. (11.28), we obtain

$$\Delta\Omega_s = \frac{\Delta\kappa}{2 \cdot |\beta_1''| \cdot \Omega_s} = \frac{\pi}{L_w \cdot |\beta_1''| \cdot \Omega_s} \tag{11.32}$$

Pulse compression exploits (group velocity) dispersion and Kerr nonlinearity. This leads to the shortest pulses ever produced: 6 fs, representing about three optical periods! As shown in Section 9.2, the Kerr effect yields an enlargement of bandwidth (self-phase modulation). For $\lambda < \lambda_D$, one has normal dispersion: $\beta_0'' > 0$, and one obtains more rapid pulse broadening. However, it is important now that the pulse gets an almost linear chirp. If this transformed pulse is sent subsequently through a pair of parallel gratings (see Section 9.1.4), which represent an optical element with negative group velocity dispersion (analogously to glass at $\lambda > \lambda_D$), one reaches, for example, the above quoted considerable pulse compression. Instead of a grating pair, one can, of course, also use a fiber, which actually shows a value of $\beta_0'' < 0$ at the wavelength under consideration.

Solitons are possibly of interest in optical communication in the future since with them extremely high-bitrate data communication over large distances can be realized without the usual bandwidth limitation due to fiber dispersion.

Real fibers possess damping. In this case, we have to add in Eq. (9.50) on the right-hand side a damping term $-j(\alpha/2) \cdot \bar{E}$, where α is the damping constant for power. For the generation of a soliton, the magnitude of this term should be smaller than the nonlinear term $G|\bar{E}|^2\bar{E} = k_0 n_2 |\bar{E}|^2\bar{E}$. For an electric field with 10^6 V/m, the nonlinear coefficient $G|\bar{E}|^2$ at a wavelength of $\lambda = 1.5\,\mu m$ has an order of magnitude of 0.2/km; the damping term $\alpha/2$ should thus be smaller. In the limit, we therefore set $\alpha/2 = 0.2$/km. The decrease in power is then $e^{-\alpha z} = 10^{-(d/10)z}$. This means that loss rates of less than $d = 1.7$ dB/km are necessary. This can easily be achieved since fibers with a damping < 0.2 dB/km are commercially available now.

If the nonlinear Schrödinger equation is now solved using perturbation theory, the solution is still Eq. (10.19) up to an additional factor $\exp(-\alpha z)$, that is, the soliton amplitude decays exponentially, but the width, determined by the argument of the new sech function, increases now exponentially as $\exp(\alpha z)$. Thus, the product amplitude × pulse width remains constant. The energy decreases due to the nonlinearity as $\exp(-2\alpha z)$, twice as rapidly as a linear pulse. In order to send solitons on long distances, amplification is necessary.

With Raman amplification, the original shape of the soliton can be restored in a real fiber. Contrary to the loss mechanism, the pulse width in soliton amplification will be reduced and the amplitude enlarged in such a manner that the product remains constant. One can adjust the Raman gain in such a way that the damping losses are just balanced and the total loss rate vanishes; this means setting the power-damping constant α to zero. Then, the soliton propagates without distortion. In an experiment, light from Raman pump sources was launched at equal distances (e.g., every 50 km) into the fiber in both directions. In this way, pulses with a bit rate of 10 Gbit/s have been transmitted over more then 10^6 km (Favre et al., 1995) and the pulse shape at the end was nearly unchanged. Moreover, one also has tried an amplification of solitons by Erbium-doped fibers. Compared with Raman amplification, a much smaller pump power is required.

As a perturbing effect, the *Gordon–Haus effect* occurs during amplification: Each single amplifier of the whole amplification chain contributes amplification noise from spontaneous emission. This shifts statistically the frequency of the soliton. (Or, the stable soliton wants to keep its shape and reacts on perturbation with a frequency shift.) Via the group velocity dispersion $dv_g/d\omega$, the group velocities change also statistically, and due to the statistically shifted times of arrival, a time jitter occurs. The perturbing Gordon–Haus effect can be reduced considerably by appropriate measures.

REFERENCES

F. Favre, D. LeGuen and M. L. Moulinard. Robustness of 20 Gbit/s 63 km span 6 mm sliding-filter controlled soliton transmission. *Electron. Lett.* 31 (1995): 1600–1601.

R. G. Smith. Optical power handling capacity of low loss optical fibers as determined by stimulated Raman and Brillouin scattering. *Appl. Opt.* 11 (1972): 2489–2496.

BIBLIOGRAPHY

G. P. Agrawal. *Nonlinear Fiber Optics.* Wiley, New York, 1984.

D. Cotter. Fibre nonlinearities in optical communications. *Opt. Quant. Elect.* 19 (1987): 1–17.

P. A. Fisher (ed.). *Optical Phase Conjugation.* Academic Press, New York, 1983.

S. E. Miller and A. G. Chynoweth (eds.). *Optical Fiber Telecommunication.* Academic Press, New York, 1979.

Y. R. Shen. *The Principles of Nonlinear Optics.* Wiley, New York, 1984.

A. E. Siegman. *Lasers,* University Science Books, Mill Valley, CA. 1986.

Notation

1. Table A.1 lists the most often used abbreviations.

TABLE A.1 Abbreviations

ADP	ammonium dihyrogen phosphate = $NH_4H_2PO_4$
c.c.	complex conjugate
FT	Fourier transform
FWHM	full width at half maximum
FWM	four-wave mixing
IBE	induced Brillouin effect
IRE	induced Raman effect
IST	inverse scattering theory
KDP	potassium dihydrogen phosphate = KH_2PO_4
KdV	Korteweg–de Vries
NLS	nonlinear Schrödinger equation
ODE	ordinary differential equation
PDE	partial differential equation
SHG	second harmonic generation
SRE	spontaneous Raman effect
SVAA	slowly varying amplitude approximation
SVEA	slowly varying envelope approximation
WDM	wavelength division multiplex

2. $Z_0 = \sqrt{\mu_0/\varepsilon_0} = (\varepsilon_0 c)^{-1} = 377$ Ohm denotes the impedance of free space.

3. The Fourier transform (FT) of a real function $G(t)$ will be written in the form

$$G_f(f) = \mathcal{F}[G(t)] = \int_{-\infty}^{\infty} G(t)e^{-2\pi jft}dt = \int_{-\infty}^{+\infty} G(t)e^{-j\omega t}dt \qquad (A.1)$$

169

Therefore,

$$G_f(-f) = G_f^*(f), \qquad G_f(f = 0) \text{ real}$$

Possibly, one should use truncated functions in order to avoid mathematical difficulties.

For a *monochromatic* signal $G(t)$ with frequency $f_1 (> 0)$, we write

$$G(t) = \text{Re } \tilde{G}(t) = \text{Re}[\widehat{G}(f_1)e^{j\omega_1 t}] = \left(\frac{1}{2}\right)[\widehat{G}(f_1)e^{j\omega_1 t} + \text{c.c.}]$$

$$= \left(\frac{1}{2}\right)\sum_{r=\pm 1}\widehat{G}(f_r)e^{j\omega_r t} = |\widehat{G}(f_1)| \cdot \cos[\omega_1 t + \arg \widehat{G}(f_1)] \qquad \text{(A.2)}$$

with the definitions

$$f_{-r} := -f_r, \qquad \widehat{G}(-f_1) := \widehat{G}^*(f_1)$$

therefore, $\widehat{G}(0)$ real.

$$\tilde{G}(t) = \widehat{G}(f_1)e^{j\omega_1 t}$$

is the analytic signal, and

$$\widehat{G}(f_1) = |\widehat{G}(f_1)| \cdot \exp[j \cdot \arg \widehat{G}(f_1)] \qquad \text{(A.3)}$$

the complex phasor. The Fourier transform of the monochromatic signal $G(t)$ is

$$G_f(f) = \frac{1}{2}[\widehat{G}(f_1) \cdot \delta(f - f_1) + \widehat{G}^*(f_1) \cdot \delta(f + f_1)] = \frac{1}{2}\sum_{r=\pm 1}\widehat{G}(f_r) \cdot \delta(f - f_r)$$

$$\text{(A.4)}$$

[*Note*: $\int \exp(\pm 2\pi juv) \, du = \delta(v)$.] The subscript f for the functions in the frequency domain will be omitted henceforth, if there is no danger of confusion. $G(t)$ and $\widehat{G}(f_1)$ possess equal dimensions, in contrast to $G(t)$ and $G_f(f)$!

As a matter of fact, all fields depend on spatial coordinates and time: $G(\mathbf{x}, t)$. We extract frequently the factor $\exp(-j\mathbf{k}\mathbf{x})$ out of the phasor:

$$\widehat{G}(\mathbf{x}, f_1) = \overline{G}(\mathbf{k}, f_1) \cdot \exp(-j\mathbf{k}\mathbf{x}) \qquad \text{(A.5)}$$

The complex function $\overline{G}(\mathbf{k}, f_1)$ will be only slowly variable with $\mathbf{x}$ or even independent of $\mathbf{x}$ (which will be specified in each case).

4. For analytic signals

$$\widetilde{E}(z,t) = \widehat{E}(z,f)e^{j\omega t} = \overline{E}(k,f)e^{j(\omega t - kz)}$$

[with Re $\widetilde{E}(z,t) = E(z,t)$ and $|\widetilde{E}(z,t)|^2 = |\widehat{E}(z,f_1)|^2 = |\overline{E}(z,f_1)|^2$], we define an averaged density of energy flow ($=$ intensity I) as the time average of the Poynting vector $\mathbf{S} = S \cdot \mathbf{e}_z$:

$$I := \langle S(z,t) \rangle = \frac{1}{2} \operatorname{Re}\, (\widetilde{E}^* \widetilde{H}) = \frac{n}{2 \cdot Z_0} |\widetilde{E}(z,t)|^2 = \frac{\mathcal{P}}{A} \qquad (A.6)$$

The dimension of $\langle S \rangle$ is (power $\mathcal{P}$)/(cross-sectional area A).

For a scalar field $\widehat{\phi}$, we define as the averaged density of energy flow

$$I = \langle S(z,t) \rangle = \left(\frac{n}{2} \right) \cdot |\widehat{\phi}|^2 \qquad (A.7)$$

Then, the dimension of $\widehat{\phi}$ is already fixed: [power/area]$^{1/2}$.

5. We use frequently the summation convention: Behind a summation symbol $\sum$, one has to sum over all those Cartesian indices that occur several times in the following term. The Cartesian components of vectors (subscripts x, y, z) are also denoted by the subscripts $1, 2, 3$.

6. The components of the electric field vector $\mathbf{E}$, the unit vector $\mathbf{e}$, and the momentum vector $\mathbf{k}$ possess sometimes two subscripts: The first of them denotes a number, the second the Cartesian component. Example: E_{ri} denotes the ith component of the rth field vector.

7. We frequently omit the superscript (n) from the susceptibility tensor $\chi^{(n)}_{ijkl...}$ and the polarization $\mathbf{P}^{(n)}$.

8. If $\mathbf{q}$ is a tensor of rank two, then $\mathbf{q} \cdot \mathbf{E}$ denotes a vector with components $\sum_{j=1}^{3} q_{ij} E_j$; if $\mathbf{q}$ is a tensor of rank three, then $\mathbf{q}{:}\mathbf{EE}$ denotes a vector with components $\sum_{j,k=1}^{3} q_{ijk} E_j E_k$.

Conversion of Units for Susceptibility

A physical quantity Z can always be written in the form

$$Z = Z_e \cdot \text{unit}, \qquad Z_e \in \Re \tag{B.1}$$

We shall convert such quantities between the SI system of units, the system we use in this book, and CGS. Quantities in SI will be written as Z, in the system of electrostatic CGS units (esu) as $\overline{Z}$. Z_e and $\overline{Z}_e$, respectively, are the numerical values. 1 esu(Q) means, for example, one electrostatic unit of charge (CGS).

Force is the same physical quantity in both systems:

$$F = F_e \cdot \text{Newton} = F_e \cdot \text{kg} \cdot \text{m/s}^2 \overset{!}{=} \overline{F}_e \cdot \text{dyn} = \overline{F}_e \cdot \text{g} \cdot \text{cm/s}^2 \tag{B.2}$$

$$1\,\text{N} = 1\,\text{W} \cdot \text{s/m} = 1\,\text{AV s/m}, 1\,\text{dyn} = 1\,\text{g} \cdot \text{cm/s}^2 = 10^{-5}\text{N} \tag{B.3}$$

The same holds for lengths: $R = \overline{R}$.

The Coulomb law yields the connection between both systems:

$$F = \frac{Q_1 Q_2}{4\pi\varepsilon_0 R^2} = \overline{F} = \frac{\overline{Q}_1 \overline{Q}_2}{\overline{R}^2} \tag{B.4}$$

Therefore, we have for equal charges $Q_1 = Q_2$ and $\overline{Q}_1 = \overline{Q}_2$, respectively:

$$Q = \sqrt{4\pi\varepsilon_0} \cdot \overline{Q} \quad (Q \neq \overline{Q}) \tag{B.5}$$

or

$$Q_e \cdot \text{As} = \sqrt{4\pi\varepsilon_0} \cdot \overline{Q}_e \cdot \text{esu}(Q) \tag{B.6}$$

$$1 \text{ esu}(Q) = 1 \text{ cm}^{3/2} \text{ g}^{1/2} \text{ s}^{-1} = 10^{-9/2} \text{ m}^{1/2} \text{ J}^{1/2} \tag{B.7}$$

[this is a consequence of the last equality in Eq. (B.4)]. With

$$\varepsilon_0 = (36\pi \cdot 10^9)^{-1} \cdot (\text{As})^2/(\text{J} \cdot \text{m}) = (36\pi \cdot 10^9)^{-1} \text{As} /(\text{V} \cdot \text{m}) \tag{B.8}$$

we find then

$$\sqrt{4\pi\varepsilon_0} = \frac{1}{3} \cdot 10^{-9/2} \cdot \text{As} \cdot (\text{J} \cdot \text{m})^{-1/2} = \frac{1}{3} \cdot 10^{-9} \cdot \text{As/esu}(Q) \tag{B.9}$$

Therefore, for the numerical values

$$Q_e = \frac{\overline{Q}_e}{(3 \cdot 10^9)} \tag{B.10}$$

With the electric field, we have

$$F = Q \cdot E = \overline{F} = \overline{Q} \cdot \overline{E} \tag{B.11}$$

and with Eq. (B.5),

$$E = E_e \cdot \text{V/m} = \overline{E} \cdot \frac{\overline{Q}}{Q} = (4\pi\varepsilon_0)^{-1/2} \cdot \overline{E} = (4\pi\varepsilon_0)^{-1/2} \cdot \overline{E}_e \cdot \text{esu}(U)/\text{cm} \tag{B.12}$$

With $1 \text{ esu}(U) = \text{g}^{1/2} \text{ cm}^{1/2} \text{ s}^{-1}$ [this follows from Eqs. (B.7) and (B.11)], we get for the numerical values

$$E_e = 3 \cdot 10^4 \cdot \overline{E}_e \tag{B.13}$$

Finally, we use the relations

$$D = \varepsilon_0 E + P \quad \text{or} \quad \overline{D} = \overline{E} + 4\pi\overline{P} \tag{B.14}$$

Then

$$\overline{D} - 4\pi\overline{P} = \overline{E} = \sqrt{4\pi\varepsilon_0} \cdot E = \sqrt{\frac{4\pi}{\varepsilon_0}} \cdot D - \sqrt{\frac{4\pi}{\varepsilon_0}} \cdot P$$

Therefore

$$D = \frac{\varepsilon_0}{\sqrt{4\pi\varepsilon_0}} \cdot \overline{D}, \quad P = \sqrt{4\pi\varepsilon_0} \cdot \overline{P} \tag{B.15}$$

In the frequency domain, [see Eq. (1.29)], we have for

$$P = \varepsilon_0 \sum \chi^{(j)} \otimes E^j, \quad \overline{P} = \sum \overline{\chi}^{(j)} \otimes \overline{E}^j \tag{B.16}$$

($\otimes$ = j-fold tensorial product, no convolution!) with Eqs. (B.12) and (B.15),

$$\chi^{(j)} = 4\pi (4\pi\varepsilon_0)^{(j-1)/2} \cdot \overline{\chi}^{(j)} \tag{B.17}$$

In particular, for the susceptibility tensors of first, second and third rank ($j = 1, 2, 3$),

$$\chi^{(1)} = 4\pi\overline{\chi}^{(1)}$$

$$\chi^{(2)} = 4\pi\sqrt{4\pi\varepsilon_0}\overline{\chi}^{(2)} = \overline{\chi}^{(2)}\frac{4\pi}{3} \cdot 10^{-9} \cdot \text{As/esu}(Q)$$

$$\chi^{(3)} = 4\pi \cdot 4\pi\varepsilon_0\overline{\chi}^{(3)} = \overline{\chi}^{(3)} \cdot \frac{4\pi}{9} \cdot 10^{-18} \cdot (\text{As})^2/\text{esu}^2(Q) \tag{B.18}$$

Introducing the units gives

$$\chi^{(j)} = \chi_e^{(j)} \cdot (\text{m/V})^{j-1}, \quad \overline{\chi}^{(j)} = \overline{\chi}_e^{(j)} \cdot [\text{cm /esu}(U)]^{j-1} \tag{B.19}$$

Then from Eq. (B.17)

$$\chi_e^{(j)} = 4\pi \cdot (3 \cdot 10^4)^{1-j} \cdot \overline{\chi}_e^{(j)}. \tag{B.20}$$

BIBLIOGRAPHY

J. D. Jackson. *Classical Electrodynamics*, (2nd ed.) J. Wiley, New York, 1975.
I. M. Skinner and S. J. Garth. Reconciliation of esu and mksa units in nonlinear optics. *Am. J. Phys.* 58 (1990): 177–181.

Tables of Susceptibility Tensors*

Following are supplements to and explanations for Tables C.1, C.2, C.3.

The set of all symmetry elements of a crystal is called its *crystal class*. There are 32 different crystal classes (or point groups). They are classified into seven *crystal systems*: triclinic, monoclinic, orthorhombic, tetragonal, hexagonal, trigonal, and cubic. The different classes are characterized by the generating symmetry elements: '*n*' denotes, for example, an *n*-fold rotation axis (a rotation of $2\pi/n$ leaves the crystal invariant), '$\bar{n}$' an *n*-fold rotation followed (or preceded) by an inversion, '*m*' a mirror plane ($= \bar{2}$), '*n/m*' an *n*-fold rotation axis with the mirror plane perpendicular to it.

Both the frequently used nonlinear crystals KDP = potassium dihydrogen phosphate = KH_2PO_4 and ADP = ammonium dihydrogen phosphate = $NH_4H_2PO_4$ belong to the crystal class $\bar{4}2\,m$ of the tetragonal system, whereas the "Banana" crystal = $Ba_2NaNb_5O_{15}$ belongs to the class *mm*2 of the orthorhombic system and lithium niobate = $LiNbO_3$ to the class 3*m* of the trigonal system. Finally, gallium arsenide = GaAs belongs to the class $\bar{4}3\,m$ of the cubic system.

REFERENCES

P. N. Butcher. *Nonlinear Optical Phenomena*. Bulletin 200, Engineering Experimental Station, Ohio State University, Columbus, Ohio, 1965.

C. C. Shang and H. Hsu. The spatial symmetric forms of third-order nonlinear suscept-ibility. *J. Quant. Elect*. IEEE-QE 23 (1987): 177–179.

---

*Reproduced with permission from (Butcher, 1965) and corrected according to (Shang and Hsu, 1987).

TABLE C.1   First-Order Susceptibility Tensors

The form of the first-order susceptibility tensor $\chi_{\mu\alpha}^{(1)}(\omega)$ for the seven crystal systems and for isotropic media. Each element is denoted only by its subscripts. The number in parenthesis is the number of independent nonzero elements.

Triclinic
$$\begin{bmatrix} xx & xy & zx \\ xy & yy & yz \\ zx & yz & zz \end{bmatrix} \tag{6}$$

Monoclinic
$$\begin{bmatrix} xx & 0 & zx \\ 0 & yy & 0 \\ zx & 0 & zz \end{bmatrix} \tag{4}$$

Orthorhombic
$$\begin{bmatrix} xx & 0 & 0 \\ 0 & yy & 0 \\ 0 & 0 & zz \end{bmatrix} \tag{3}$$

Tetragonal
Trigonal
Hexagonal
$$\begin{bmatrix} xx & 0 & 0 \\ 0 & xx & 0 \\ 0 & 0 & zz \end{bmatrix} \tag{2}$$

Cubic
Isotropic
$$\begin{bmatrix} xx & 0 & 0 \\ 0 & xx & 0 \\ 0 & 0 & xx \end{bmatrix} \tag{1}$$

TABLE C.2   Second-Order Susceptibility Tensors

The form of the second-order susceptibility tensor $\chi_{\mu\alpha\beta}^{(2)}(\omega_1,\omega_2,)$ for those crystal classes which have no center of symmetry. Each element is denoted only by its subscripts and a bar denotes the negative. The number in parenthesis is the number of independent nonzero elements. The tensor is identically zero for those classes which do not appear.

Triclinic

Class 1
$$\begin{bmatrix} xxx & xyy & xzz & xyz & xzy & xzx & xxz & xxy & xyx \\ yxx & yyy & yzz & yyz & yzy & yzx & yxz & yxy & yyx \\ zxx & zyy & zzz & zyz & zzy & zzx & zxz & zxy & zyx \end{bmatrix} \tag{27}$$

Monoclinic

Class 2
$$\begin{bmatrix} 0 & 0 & 0 & xyz & xzy & 0 & 0 & xxy & xyx \\ yxx & yyy & yzz & 0 & 0 & yzx & yxz & 0 & 0 \\ 0 & 0 & 0 & zyz & zzy & 0 & 0 & zxy & zyx \end{bmatrix} \tag{13}$$

Class m
$$\begin{bmatrix} xxx & xyy & xzz & 0 & 0 & xzx & xxz & 0 & 0 \\ 0 & 0 & 0 & yyz & yzy & 0 & 0 & yxy & yyx \\ zxx & zyy & zzz & 0 & 0 & zzx & zxz & 0 & 0 \end{bmatrix} \tag{14}$$

Orthorhombic

Class 222

$$\begin{bmatrix} 0 & 0 & 0 & xyz & xzy & 0 & 0 & 0 & 0 \\ 0 & 0 & 0 & 0 & 0 & yzx & yxz & 0 & 0 \\ 0 & 0 & 0 & 0 & 0 & 0 & 0 & zxy & zyx \end{bmatrix} \tag{6}$$

Class mm2

$$\begin{bmatrix} 0 & 0 & 0 & 0 & 0 & xzx & xxz & 0 & 0 \\ 0 & 0 & 0 & yyz & yzy & 0 & 0 & 0 & 0 \\ zxx & zyy & zzz & 0 & 0 & 0 & 0 & 0 & 0 \end{bmatrix} \tag{7}$$

Tetragonal

Class 4

$$\begin{bmatrix} 0 & 0 & 0 & xyz & xzy & xzx & xxz & 0 & 0 \\ 0 & 0 & 0 & xxz & xzx & \overline{xzy} & \overline{xyz} & 0 & 0 \\ zxx & zxx & zzz & 0 & 0 & 0 & 0 & zxy & \overline{zxy} \end{bmatrix} \tag{7}$$

Class $\bar{4}$

$$\begin{bmatrix} 0 & 0 & 0 & xyz & xzy & xzx & xxz & 0 & 0 \\ 0 & 0 & 0 & \overline{xxz} & \overline{xzx} & xzy & xyz & 0 & 0 \\ zxx & \overline{zxx} & 0 & 0 & 0 & 0 & 0 & zxy & zxy \end{bmatrix} \tag{6}$$

Class 422

$$\begin{bmatrix} 0 & 0 & 0 & xyz & xzy & 0 & 0 & 0 & 0 \\ 0 & 0 & 0 & 0 & 0 & \overline{xzy} & \overline{xyz} & 0 & 0 \\ 0 & 0 & 0 & 0 & 0 & 0 & 0 & zxy & \overline{zxy} \end{bmatrix} \tag{3}$$

Class 4mm

$$\begin{bmatrix} 0 & 0 & 0 & 0 & 0 & xzx & xxz & 0 & 0 \\ 0 & 0 & 0 & xxz & xzx & 0 & 0 & 0 & 0 \\ zxx & zxx & zzz & 0 & 0 & 0 & 0 & 0 & 0 \end{bmatrix} \tag{4}$$

Class $\overline{4}$2m

$$\begin{bmatrix} 0 & 0 & 0 & xyz & xzy & 0 & 0 & 0 & 0 \\ 0 & 0 & 0 & 0 & 0 & xzy & xyz & 0 & 0 \\ 0 & 0 & 0 & 0 & 0 & 0 & 0 & zxy & zxy \end{bmatrix} \tag{3}$$

Cubic

Class 432

$$\begin{bmatrix} 0 & 0 & 0 & xyz & \overline{xyz} & 0 & 0 & 0 & 0 \\ 0 & 0 & 0 & 0 & 0 & xyz & \overline{xyz} & 0 & 0 \\ 0 & 0 & 0 & 0 & 0 & 0 & 0 & xyz & \overline{xyz} \end{bmatrix} \tag{1}$$

Class $\overline{4}$3m

$$\begin{bmatrix} 0 & 0 & 0 & xyz & xyz & 0 & 0 & 0 & 0 \\ 0 & 0 & 0 & 0 & 0 & xyz & xyz & 0 & 0 \\ 0 & 0 & 0 & 0 & 0 & 0 & 0 & xyz & xyz \end{bmatrix} \tag{1}$$

Class 23

$$\begin{bmatrix} 0 & 0 & 0 & xyz & xzy & 0 & 0 & 0 & 0 \\ 0 & 0 & 0 & 0 & 0 & xyz & xzy & 0 & 0 \\ 0 & 0 & 0 & 0 & 0 & 0 & 0 & xyz & xzy \end{bmatrix} \tag{2}$$

Trigonal

Class 3

$$\begin{bmatrix} xxx & \overline{xxx} & 0 & xyz & xzy & xzx & xxz & \overline{yyy} & \overline{yyy} \\ \overline{yyy} & yyy & 0 & xxz & xzx & \overline{xzy} & \overline{xyz} & \overline{xxx} & \overline{xxx} \\ zxx & zxx & zzz & 0 & 0 & 0 & 0 & zxy & \overline{zxy} \end{bmatrix} \quad (9)$$

Class 32

$$\begin{bmatrix} xxx & \overline{xxx} & 0 & xyz & xzy & 0 & 0 & 0 & 0 \\ 0 & 0 & 0 & 0 & 0 & \overline{xzy} & \overline{xyz} & \overline{xxx} & \overline{xxx} \\ 0 & 0 & 0 & 0 & 0 & 0 & 0 & zxy & \overline{zxy} \end{bmatrix} \quad (4)$$

Class 3m

$$\begin{bmatrix} 0 & 0 & 0 & 0 & 0 & xzx & xxz & \overline{yyy} & \overline{yyy} \\ \overline{yyy} & yyy & 0 & xxz & xzx & 0 & 0 & 0 & 0 \\ zxx & zxx & zzz & 0 & 0 & 0 & 0 & 0 & 0 \end{bmatrix} \quad (5)$$

Hexagonal

Class 6

$$\begin{bmatrix} 0 & 0 & 0 & xyz & xzy & xzx & xxz & 0 & 0 \\ 0 & 0 & 0 & xxz & xzx & \overline{xzy} & \overline{xyz} & 0 & 0 \\ zxx & zxx & zzz & 0 & 0 & 0 & 0 & zxy & \overline{zxy} \end{bmatrix} \quad (7)$$

Class $\bar{6}$

$$\begin{bmatrix} xxx & \overline{xxx} & 0 & 0 & 0 & 0 & 0 & \overline{yyy} & \overline{yyy} \\ \overline{yyy} & yyy & 0 & 0 & 0 & 0 & 0 & \overline{xxx} & \overline{xxx} \\ 0 & 0 & 0 & 0 & 0 & 0 & 0 & 0 & 0 \end{bmatrix} \quad (2)$$

Class 622

$$\begin{bmatrix} 0 & 0 & 0 & xyz & xzy & 0 & 0 & 0 & 0 \\ 0 & 0 & 0 & 0 & 0 & \overline{xzy} & \overline{xyz} & 0 & 0 \\ 0 & 0 & 0 & 0 & 0 & 0 & 0 & zxy & \overline{zxy} \end{bmatrix} \quad (3)$$

Class 6mm

$$\begin{bmatrix} 0 & 0 & 0 & 0 & 0 & xzx & xxz & 0 & 0 \\ 0 & 0 & 0 & xxz & xzx & 0 & 0 & 0 & 0 \\ zxx & zxx & zzz & 0 & 0 & 0 & 0 & 0 & 0 \end{bmatrix} \quad (4)$$

Class $\bar{6}$ m 2

$$\begin{bmatrix} 0 & 0 & 0 & 0 & 0 & 0 & 0 & \overline{yyy} & \overline{yyy} \\ \overline{yyy} & yyy & 0 & 0 & 0 & 0 & 0 & 0 & 0 \\ 0 & 0 & 0 & 0 & 0 & 0 & 0 & 0 & 0 \end{bmatrix} \quad (1)$$

TABLE C.3   Third-Order Susceptibility Tensors
The form of the third-order susceptibility tensor $\chi^{(3)}_{\mu\alpha\beta\gamma}(\omega_1, \omega_2, \omega_3)$ for the 32 crystal classes and isotropic media. Each element is denoted only by its subscripts and a bar denotes the negative.

Triclinic
For both classes, 1 and $\bar{1}$, there are 81 independent nonzero elements.

Monoclinic
For all three classes, 2, m, and 2/m, there are 41 independent nonzero elements, consisting of:

  3 elements with suffixes all equal
  18 elements with suffixes equal in pairs
  12 elements with suffixes having two y's, one x, and one z

4 elements with suffixes having three x's and one z
4 elements with suffixes having three z's and one x.

Orthorhombic

For all three classes, 222, mm2, and mmm, there are 21 independent nonzero elements, consisting of:

    3 elements with suffixes all equal
    18 elements with suffixes equal in pairs

Tetragonal

For the three classes 4, $\bar{4}$, and 4/m, there are 41 nonzero elements of which only 21 are independent. They are:

$$xxxx = yyyy \qquad zzzz$$

$zzxx$	$=$	$zzyy$	$xyzz$	$=$	$\overline{yxzz}$	$xxyy$	$=$	$yyxx$	$xxxy = \overline{yyyx}$
$xxzz$	$=$	$yyzz$	$zzxy$	$=$	$\overline{zzyx}$	$xyxy$	$=$	$yxyx$	$xxyx = \overline{yyxy}$
$zxzx$	$=$	$zyzy$	$xzyz$	$=$	$\overline{yzxz}$	$xyyx$	$=$	$yxxy$	$xyxx = \overline{yxyy}$

$xzxz$	$=$	$yzyz$	$zxzy$	$=$	$\overline{zyzx}$		$yxxx = \overline{xyyy}$
$zxxz$	$=$	$zyyz$	$zxyz$	$=$	$\overline{zyxz}$		
$xzzx$	$=$	$yzzy$	$xzzy$	$=$	$\overline{yzzx}$		

For the four classes 422, 4mm, 4/mmm, and $\bar{4}$2m, there are 21 nonzero elements of which only 11 are independent. They are:

$$xxxx = yyyy \qquad zzzz$$

$yyzz$	$=$	$xxzz$	$yzzy$	$=$	$xzzx$	$xxyy$	$=$	$yyxx$
$zzyy$	$=$	$zzxx$	$yzyz$	$=$	$xzxz$	$xyxy$	$=$	$yxyx$
$zyyz$	$=$	$zxxz$	$zyzy$	$=$	$zxzx$	$xyyx$	$=$	$yxxy$

Cubic

For the two classes 23 and m3, there are 21 nonzero elements of which only 7 are independent. They are:

$xxxx$	$=$	$yyyy$	$=$	$zzzz$
$yyzz$	$=$	$zzxx$	$=$	$xxyy$
$zzyy$	$=$	$xxzz$	$=$	$yyxx$
$yzyz$	$=$	$zxzx$	$=$	$xyxy$
$zyzy$	$=$	$xzxz$	$=$	$yxyx$
$yzzy$	$=$	$zxxz$	$=$	$xyyx$
$zyyz$	$=$	$xzzx$	$=$	$yxxy$

For the three classes 432, $\bar{4}$3m, and m3m, there are 21 nonzero elements of which only 4 are independent. They are:

$xxxx$	$=$	$yyyy$	$=$	$zzzz$							
$yyzz$	$=$	$zzyy$	$=$	$zzxx$	$=$	$xxzz$	$=$	$xxyy$	$=$	$yyxx$
$yzyz$	$=$	$zyzy$	$=$	$zxzx$	$=$	$xzxz$	$=$	$xyxy$	$=$	$yxyx$
$yzzy$	$=$	$zyyz$	$=$	$zxxz$	$=$	$xzzx$	$=$	$xyyx$	$=$	$yxxy$

Trigonal

For the two classes 3 and $\bar{3}$, there are 73 nonzero elements of which only 27 are independent. They are:

$$zzzz$$
$$xxxx = yyyy = xxyy + xyyx + xyxy \begin{cases} xxyy = yyxx \\ xyyx = yxxy \\ xyxy = yxyx \end{cases}$$

$$
\begin{aligned}
yyzz &= xxzz & xyzz &= \overline{yxzz} \\
zzyy &= zzxx & zzxy &= \overline{zzyx} \\
zyyz &= zxxz & zxyz &= \overline{zyxz} \\
yzzy &= xzzx & xzzy &= \overline{yzzx} \\
yzyz &= xzxz & xzyz &= \overline{yzxz} \\
zyzy &= zxzx & zxzy &= \overline{zyzx}
\end{aligned}
$$

$$xxyy = \overline{yyyx} = yyxy + yxyy + xyyy \begin{cases} yyxy = \overline{xxyx} \\ yxyy = \overline{xyxx} \\ xyyy = \overline{yxxx} \end{cases}$$

$$
\begin{aligned}
yyyz &= \overline{yxxz} &= \overline{xyxz} &= \overline{xxyz} \\
yyzy &= \overline{yxzx} &= \overline{xyzx} &= \overline{xxzy} \\
yzyy &= \overline{yzxx} &= \overline{xzyz} &= \overline{xzxy} \\
zyyy &= \overline{zyxx} &= \overline{zxyx} &= \overline{zxxy} \\
xxxz &= \overline{xyyz} &= \overline{yxyz} &= \overline{yyxz} \\
xxzx &= \overline{xyzy} &= \overline{yxzy} &= \overline{yyzx} \\
xzxx &= \overline{yzxy} &= \overline{xzyy} &= \overline{yzyx} \\
zxxx &= \overline{zxyy} &= \overline{zyxy} &= \overline{zyyx}
\end{aligned}
$$

For the three classes 3m, $\bar{3}$m, and 32 there are 37 nonzero elements of which only 14 are independent. They are:

$$zzzz$$
$$xxxx = yyyy = xxyy + xyyx + xyxy \begin{cases} xxyy = yyxx \\ xyyx = yxxy \\ xyxy = yxyx \end{cases}$$

$$
\begin{aligned}
yyzz &= xxzz & xxxz &= \overline{xyyz} &= \overline{yxyz} &= \overline{yyxz} \\
zzyy &= zzxx & xxzx &= \overline{xyzy} &= \overline{yxzy} &= \overline{yyzx} \\
zyyz &= zxxz & xzxx &= \overline{xzyy} &= \overline{yzxy} &= \overline{yzyx} \\
yzzy &= xzzx & zxxx &= \overline{zxyy} &= \overline{zyxy} &= \overline{zyyx} \\
yzyz &= xzxz \\
zyzy &= zxzx
\end{aligned}
$$

Hexogonal

For the three classes 6, $\bar{6}$, and 6/m, there are 41 nonzero elements of which only 19 are independent. They are:

$$zzzz$$
$$xxxx = yyyy = xxyy + xyyx + xyxy \begin{cases} xxyy = yyxx \\ xyyx = yxxy \\ xyxy = yxyx \end{cases}$$

$$
\begin{aligned}
yyzz &= xxzz & xyzz &= \overline{yxzz} \\
zzyy &= zzxx & zzxy &= \overline{zzyx} \\
zyyz &= zxxz & zxyz &= \overline{zyxz} \\
yzzy &= xzzx & xzzy &= \overline{yzzx} \\
yzyz &= xzxz & xzyz &= \overline{yzxz} \\
zyzy &= zxzx & zxzy &= \overline{zyzx}
\end{aligned}
$$

$$
xxxy = \overline{yyyx} = yyxy + yxyy + xyyy
\begin{cases}
yyxy &= \overline{xxyx} \\
yxyy &= \overline{xyxx} \\
xyyy &= \overline{yxxx}
\end{cases}
$$

For the four classes 622, 6mm, 6/mmm, and $\bar{6}$m2, there are 21 nonzero elements of which only 10 are independent. They are:

$$
\begin{aligned}
&zzzz \\
&xxxx = yyyy = xxyy + xyyx + xyxy
\begin{cases}
xxyy &= yyxx \\
xyyx &= yxxy \\
xyxy &= yxyx
\end{cases}
\end{aligned}
$$

$$
\begin{aligned}
yyzz &= xxzz \\
zzyy &= zzxx \\
zyyz &= zxxz \\
yzzy &= xzzx \\
yzyz &= xzxz \\
zyzy &= zxzx
\end{aligned}
$$

Isotropic Media
There are 21 nonzero elements of which only 3 are independent. They are:

$$
\begin{aligned}
xxxx &= yyyy = zzzz \\
yyzz &= zzyy = zzxx = xxzz = xxyy = yyxx \\
yzyz &= zyzy = zxzx = xzxz = xyxy = yxyx \\
yzzy &= zyyz = zxxz = xzzx = xyyx = yxxy \\
xxxx &= xxyy + xyxy + xyyx
\end{aligned}
$$

Permutation Symmetry for Losslessness*

Theorem   In the case of losslessness, one has

$$\chi_{ijk}(-f_3, f_1, f_2) = \chi_{jik}(f_1, -f_3, f_2) \tag{D.1}$$

Proof   The mean loss power is [see Eq. (2.3)]

$$p_\ell = \lim_{T\to\infty} \frac{1}{2T} \int\limits_{-T}^{+T} \mathbf{E}(t) \cdot \frac{\partial \mathbf{P}}{\partial t} \, dt$$

$$- \lim_{T\to\infty} \frac{1}{2T} \int\limits_{-T}^{+T} dt \int\limits_{-\infty}^{+\infty} df_1 \mathbf{E}(f_1) e^{j\omega_1 t} \cdot \frac{\partial}{\partial t} \int\limits_{-\infty}^{+\infty} df_2 \mathbf{P}(f_2) e^{j\omega_2 t}$$

$\partial/\partial t$ yields an additional factor $j\omega_2$. Then we integrate over t, resulting in a Dirac delta function $\delta(f_1 + f_2)$; subsequently, we integrate over f_1. Then we obtain with Eq. (1.23)

$$p_\ell = \int\limits_{-\infty}^{+\infty} j\omega_2 \cdot \lim \sum_i \frac{E_i(-f_2) P_i(f_2)}{2T} \cdot df_2$$

$$= \varepsilon_0 \iint df_1 df_2 (j\omega_2)$$

$$\times \lim \frac{1}{2T} \sum \chi_{ijk}(-f_2, f_1, f_2 - f_1) E_i(-f_2) E_j(f_1) E_k(f_2 - f_1) \tag{D.2}$$

* Reproduced by permission of Vieweg-Verlag, Braunschweig, from G. K. Grau: *Quantenelektronik*, 1978.

If we now set $j\omega_2 = j[\omega_1 + (\omega_2 - \omega_1)]$, then p_ℓ is divided into two parts, where the first part (I_1) contains $j\omega_1$, and the second part (I_2) $j(\omega_2 - \omega_1)$. At first, we transform the second integral:

$$
I_2 = \varepsilon_0 \iint df_1 df_2 \, j(\omega_2 - \omega_1)
$$

$$
\times \lim \frac{1}{2T} \sum \chi_{ijk}(-f_2, f_1, f_2 - f_1) \cdot E_i(-f_2) E_j(f_1) E_k(f_2 - f_1)
$$

We introduce new integration variables f_1', f_2' by $f_1' = f_2$, $f_2' = f_2 - f_1$. Thus $df_1 \, df_2 = df_1' \, df_2'$. Omit the primes and exchange $f_2 \leftrightarrow f_1$:

$$
I_2 = \varepsilon_0 \iint df_1 \, df_2 \, j\omega_1
$$

$$
\times \lim \frac{1}{2T} \sum \chi_{ijk}(-f_2, f_2 - f_1, f_1) \cdot E_i(-f_2) E_j(f_2 - f_1) E_k(f_1)
$$

Next, we exchange the summation indices j and k, use the always valid Eq. (1.32), and then immediately obtain $I_2 = I_1(j\omega_1) = p_\ell/2$.

In the first integral [see Eq. (D.2)],

$$
I_1 = \frac{p_\ell}{2}
$$

$$
= \varepsilon_0 \iint df_1 \, df_2 \, j\omega_1
$$

$$
\times \lim \frac{1}{2T} \sum \chi_{ijk}(-f_2, f_1, f_2 - f_1) \cdot E_i(-f_2) E_j(f_1) E_k(f_2 - f_1)
$$

we exchange the summation indices i and j and substitute $f_2 \rightarrow -f_1, f_1 \rightarrow -f_2$. Thus,

$$
\frac{p_\ell}{2} = - \varepsilon_0 \iint df_1 \, df_2 \, j\omega_2
$$

$$
\times \lim \frac{1}{2T} \sum \chi_{jik}(f_1, -f_2, f_2 - f_1) \cdot E_i(-f_2) E_j(f_1) E_k(f_2 - f_1)
$$

If this is added to Eq. (D.2), then

$$
p_\ell + \frac{p_\ell}{2} = \frac{3p_\ell}{2} = \varepsilon_0 \iint df_1 \, df_2 \, j\omega_2 \lim \frac{1}{2T} \sum [\chi_{ijk}(-f_2, f_1, f_2 - f_1)
$$

$$
- \chi_{jik}(f_1, -f_2, f_2 - f_1)] E_i(-f_2) E_j(f_1) E_k(f_2 - f_1)
$$

results. However, in the lossless case, we have $p_\ell = 0$; thus,

$$
\chi_{ijk}(-f_2, f_1, f_2 - f_1) = \chi_{jik}(f_1, -f_2, f_2 - f_1)
$$

∎

Inverse Scattering Theory

For the solution of the nonlinear Schrödinger equation (NLS), a method is applicable that has been devised for nonlinear evolution equations with given initial values. An evolution equation for a function $u(\mathbf{x}, t)$ may be written in the form

$$\left(\frac{\partial u}{\partial t} \equiv \right) u_t = K(u), \qquad u(\mathbf{x}, 0) = u_0(\mathbf{x}) \tag{E.1}$$

where $K(\quad)$ denotes a nonlinear differential operator in $\mathbf{x}$, independent of t, and u vanishes sufficiently rapidly as $|\mathbf{x}| \to \infty$. The interesting point is that certain nonlinear problems can be solved by solving only linear problems. For a more detailed description of the inverse scattering theory, we refer to the specialized literature. Therefore, we are careless with respect to mathematical assumptions. They should be always chosen in such a way that all integrations, limits, and so on can be performed. We shall restrict our considerations to not only the NLS, but also consider other nonlinear equations. This allows us to follow the development of the method of inverse scattering.

E.1 TRANSFORMATIONS OF NONLINEAR EQUATIONS

We shall consider a series of differential equations and show how linear methods can be used for the solution of the nonlinear differential equations.

Suppose we have a linear evolution problem described by the linear PDE:

$$u_t(x, t) = j \cdot \omega\left(j \frac{\partial}{\partial x}\right) \cdot u(x, t) \equiv K(u), \qquad -\infty < x < \infty, \qquad t \geq 0 \tag{E.2}$$

where $\omega(z)$ is, say, a polynomial, and the initial condition is $u(x, 0) = u_0(x)$.

For solving this problem, we use a direct and an inverse Fourier transform

(FT) with respect to x. We assume the usual conditions for u in order that the Fourier transforms exist:

$$u(k,t) = \int_{-\infty}^{+\infty} u(x,t)e^{jkx}\,dx \tag{E.3}$$

The direct FT of Eq. (E.2) yields the ODE

$$u_t(k,t) = j \cdot \omega(k) \cdot u(k,t)$$

which can immediately be integrated to give

$$u(k,t) = u(k,0) \cdot e^{j\omega(k)t} \tag{E.4}$$

With the inverse FT, we find

$$u(x,t) = \int_{-\infty}^{+\infty} u(k,t)e^{-jkx} \cdot \frac{dk}{2\pi} \tag{E.5}$$

Therefore, the method of solution requires three steps (see Fig. E.1a):

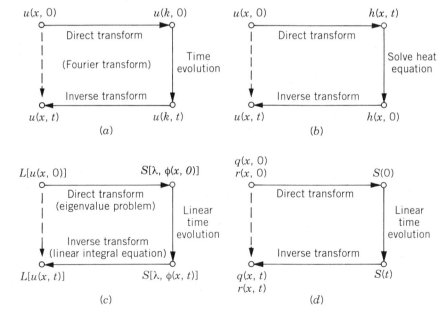

FIGURE E.1 The three linear steps for the solution of nonlinear equations: (a) linear Fourier transform, (b) Burgers equation, (c) starting with a Lax pair and (d) general matrix formalism.

1. Given the initial value $u_0(x)$, direct FT using Eq. (E.3) with $t = 0$ yields $u(k, 0)$.
2. Calculate $u(k, t)$ according to Eq. (E.4).
3. From an inverse FT, we obtain $u(x, t)$ according to Eq. (E.5).

This is, by the way, equivalent to the procedure in Section 9.1, where we calculated $\tilde{E}(z, t)$ given $\tilde{E}(0, t)$.

The second equation is the well-known Riccati differential equation

$$y_x(x) = u(x) - \lambda - y^2 \tag{E.6}$$

where λ is a constant. This nonlinear ODE (no evolution equation) can be linearized using the transformation

$$y = \frac{\psi_x}{\psi} \tag{E.7}$$

leading to

$$L\psi = \lambda\psi, \quad L = -\partial_x^2 + u \tag{E.8}$$

This is a linear, time-independent Schrödinger equation, representing an eigenvalue problem for the eigenfunction $\psi(x)$ and the constant eigenvalue λ for a given potential $u(x)$.

The third equation is a nonlinear equation called the Burgers equation

$$u_t = au_{xx} - uu_x = (a\partial_x^2 - u\partial_x)(u) \equiv K(u) \tag{E.9}$$

with $a = $ const. As the Riccati differential equation, this PDE can be linearized by a transform similar to Eq. (E.7):

$$u(x, t) = -2a \cdot \frac{h_x(x, t)}{h(x, t)} \tag{E.10}$$

(Hopf–Cole transformation) leading to a linear PDE

$$h_t(x, t) = ah_{xx}(x, t) \tag{E.11}$$

and this is the well-known heat equation. Thus to solve the initial value problem for the Burgers equation, one has to complete the following three steps (see Fig. E.1b):

1. Calculate $h(x, 0)$ from Eq. (E.10) for $t = 0$ given the initial value $u(x, 0)$:

$$h_x(x, 0) = -u(x, 0) \cdot \frac{h(x, 0)}{(2a)}$$

This direct transformation yields

$$h(x,0) = \exp\left[-\frac{1}{2a}\int u(x',0)\,dx'\right]$$

2. Solve the heat equation (E.11) for $h(x,t)$ with an initial value $h(x,0)$ using, for example, a Fourier transform.
3. The inverse (compared with step 1) transformation Eq. (E.10) yields then $u(x,t)$.

The remarkable fact is that in all three steps, we have to solve only linear problems!

We come back to the Schrödinger equation (E.8). If the potential u depends on an additional parameter t, then ψ and λ vary with t: $\psi = \psi(x,t)$, $\lambda = \lambda(t)$. If, however, the potential $u(x,t)$ is a solution of a certain nonlinear differential equation, the eigenvalues λ of Eq. (E.8) are time-independent, as we will show in the following theorem. The eigenvalues can therefore be determined from the potential $u(x,0) = u_0(x)$ for all times.

Sometimes, the evolution equation $u_t = K(u)$ can be written in the form

$$L_t = [M,L] =: ML - LM \tag{E.12}$$

where both M and L are linear operators and M is independent of t. M and L represent a so-called Lax pair and Eq. (E.12) is called the Lax equation.

Example E.1   Given $u_t = u_x$, then choose

$$L = -\partial_x^2 + u, \qquad M = \partial_x$$

Thus,

$$[M,L]\psi = u_x\psi, \qquad L_t = u_t \qquad \blacksquare$$

Example E.2   Given the standard Korteweg–de Vries equation (KdV),

$$u_t = 6uu_x - u_{xxx} \tag{E.13}$$

Here,

$$L = -\partial_x^2 + u, \qquad M = -4\partial_x^3 + 6u\partial_x + 3u_x$$

Then

$$[M,L]\psi = (6uu_x - u_{xxx})\psi, \qquad L_t = u_t \qquad \blacksquare$$

However, finding a Lax pair for a given nonlinear equation is usually a highly nontrivial task!

Theorem   If

$$L_t + [L, M] = 0$$

and L is a self-adjoint operator, then the spectrum λ from

$$L\Phi(x, t) = \lambda\Phi(x, t), \qquad -\infty < x, t < \infty \qquad (E.14)$$

is independent of t.

Proof   Differentiating Eq. (E.14) with respect to t gives

$$\lambda_t\Phi = L_t\Phi + L\Phi_t - \lambda\Phi_t = (ML - LM)\Phi + (L - \lambda)\Phi_t$$

$$= \lambda M\Phi - LM\Phi + (L - \lambda)\Phi_t = (L - \lambda)(\Phi_t - M\Phi) \qquad (E.15)$$

We now calculate the inner product defined by

$$\langle\Phi|\Phi\rangle = \int\limits_{-\infty}^{+\infty} \Phi^*\Phi \, dx > 0$$

(since Φ is an eigenfunction)

$$\lambda_t\langle\Phi|\Phi\rangle = \langle\Phi|(L - \lambda)(\Phi_t - M\Phi)\rangle$$

$$= \langle(L - \lambda)\Phi|\Phi_t - M\Phi\rangle = \langle 0|\Phi_t - M\Phi\rangle = 0$$

Hence, we must have $\lambda_t = 0$, $\lambda(t) = \lambda_0 = $ const.   ∎

The time evolution of the eigenfunctions follows then from Eq. (E.15) as

$$\Phi_t = M\Phi \qquad (E.16)$$

If L is not self-adjoint, we simply require Eq. (E.16) and $\lambda_t = 0$ follows too.

Now, the solution of the nonlinear initial value problem proceeds with the following three steps if a Lax pair has been found (see Fig. E.1c):

1. Solve the linear eigenvalue problem Eq. (E.14) for $t = 0$ with the initial value $u(x, 0)$, contained in L, yielding λ and $\Phi(x, 0)$. These are the scattering data $S(t = 0)$. This is the direct transformation.
2. Calculate $\Phi(x, t)$ using the linear ODE (E.16) with $\Phi(x, 0)$ given, yielding the scattering data $S(t)$.
3. From the scattering data $S(t)$, that is, from $\Phi(x, t)$ and λ, determine $u(x, t)$ contained in L for $t > 0$. This inverse problem requires the solution of a linear integral equation that will be derived later on.

For some nonlinear equations, the method must be generalized. This formulation is now as general as to comprise many nonlinear equations. (However, this formulation is not in all cases the most convenient one for the solution of nonlinear equations.) Consider the two linear matrix differential equations (2 × 2-matrices for simplicity):

$$\begin{bmatrix} v_1 \\ v_2 \end{bmatrix}_x = \begin{bmatrix} j\zeta & q \\ r & -j\zeta \end{bmatrix} \begin{bmatrix} v_1 \\ v_2 \end{bmatrix}, \qquad \mathbf{v}_x = \mathbf{F} \cdot \mathbf{v} \qquad (E.17)$$

$$\begin{bmatrix} v_1 \\ v_2 \end{bmatrix}_t = \begin{bmatrix} A & B \\ C & D \end{bmatrix} \begin{bmatrix} v_1 \\ v_2 \end{bmatrix} = \begin{bmatrix} A & B \\ C & -A \end{bmatrix} \begin{bmatrix} v_1 \\ v_2 \end{bmatrix}, \qquad \mathbf{v}_t = \mathbf{G} \cdot \mathbf{v} \qquad (E.18)$$

since it is sufficient for our purposes to set $D = -A$. Here, $q(x, t)$ and $r(x, t)$ are a pair of potential functions, ζ is a constant spectral parameter and A, B, C are functions of x, t, q, r, of their x-derivatives and ζ. The functions q and r will be specialized afterward in such a way that the Eqs. (E.17) and (E.18) represent some nonlinear evolution equations. For simplicity, we also assume that all eigenvalues ζ are simple. From Eqs. (E.17) and (E.18), it follows with $\mathbf{v}_{xt} = \mathbf{v}_{tx}$:

$$\mathbf{F}_t - \mathbf{G}_x + [\mathbf{F}, \mathbf{G}] = 0 \qquad (E.19)$$

or explicitly,

$$q_t = B_x - 2j\zeta B + 2qA$$
$$r_t = C_x + 2j\zeta C - 2rA \qquad (E.20)$$
$$A_x = qC - rB$$

Therefore, if $\mathbf{F}$ is given, then $\mathbf{G}$ (i.e., A, B, C) can be calculated. Of course, Eq. (E.19) is a generalization of Eq. (E.12); for $r = -1$, one finds from Eq. (E.17) the linear Schrödinger equation

$$\partial_x^2 v_2 + (\zeta^2 + q)v_2 = 0$$

[with $q(x, t) \Rightarrow -u(x, t)$, $\zeta^2 \Rightarrow \lambda$].

In order to solve Eq. (E.20), we substitute various expressions for A, B, C, with coefficients depending on x.

1. First, we use a polynomial up to second order:

$$A = A_2\zeta^2 + A_1\zeta + A_0$$
$$B = B_2\zeta^2 + B_1\zeta + B_0 \qquad (E.21)$$
$$C = C_2\zeta^2 + C_1\zeta + C_0$$

Substituting into Eq. (E.20) and equating equal powers of ζ yield

$$A = a_2\zeta^2 \qquad + \frac{1}{2}a_2qr$$

$$B = \qquad -ja_2q\zeta - \frac{1}{2}a_2q_x \qquad\qquad \text{(E.22)}$$

$$C = \qquad -ja_2r\zeta + \frac{1}{2}a_2r_x$$

where we have set $A_1 = \text{const} = 0$ and a_2 is a constant of integration. This is sufficient for our purposes. Then, the first two Eqs. (E.20) take the form

$$-\frac{1}{2}a_2q_{xx} = q_t - a_2q^2r$$

$$\frac{1}{2}a_2r_{xx} = r_t + a_2qr^2 \qquad\qquad \text{(E.23)}$$

Specializing to $r = \mp q^*, a_2 = 2j$ shows that both equations coincide and we get the nonlinear Schrödinger equation (NLS)

$$jq_t(x, t) = q_{xx}(x, t) \pm 2|q(x, t)|^2 q(x, t) \qquad\qquad \text{(E.24)}$$

2. Using polynomials up to third power in ζ for A, B, C, we find from Eq. (E.20)

$$A = a_3\zeta^3 + a_2\zeta^2 + \frac{1}{2}(a_3qr + 2a_1)\zeta + \frac{a_3}{2}qr + j\frac{a_2}{4}(qr_x - q_xr) + a_0$$
$$B = -ja_3q\zeta^2 - (ja_2q + \frac{a_3}{2}q_x)\zeta + (-ja_1q - j\frac{a_3}{2}q^2r - \frac{a_2}{2}q_x + j\frac{a_3}{4}q_{xx}) \qquad \text{(E.25)}$$
$$C = -ja_3r\zeta^2 + (-ja_2r + \frac{a_3}{2}r_x)\zeta + (-ja_1r - j\frac{a_3}{2}qr^2 + \frac{a_2}{2}r_x + j\frac{a_3}{4}r_{xx})$$

and the evolution equations

$$q_t - \frac{j}{4}a_3(q_{xxx} - 6qrq_x) + \frac{a_2}{2}(q_{xx} - 2q^2r) + ja_1q_x - 2a_0q = 0$$
$$r_t - \frac{j}{4}a_3(r_{xxx} - 6qrr_x) - \frac{a_2}{2}(r_{xx} - 2qr^2) + ja_1r_x + 2a_0r = 0 \qquad \text{(E.26)}$$

Now, we choose special values for the coefficients: $a_0 = a_1 = a_2 = 0$, $a_3 = 4j, r = 1$ again gives the KdV equation (E.13); $a_0 = a_1 = a_2 = 0, a_3 = 4j$, $r = \mp q$ yields the modified KdV equation

$$q_t \pm 6q^2q_x + q_{xxx} = 0$$

$a_0 = a_1 = a_3 = 0, a_2 = 2j, r = \mp q^*$ again gives the NLS Eq. (E.24).

3. However, if we set

$$A = \frac{a(x,t)}{\zeta}, \qquad B = \frac{b(x,t)}{\zeta}, \qquad C = \frac{c(x,t)}{\zeta} \qquad \text{(E.27)}$$

we get

$$a_x = -\frac{j}{2}(qr)_t, \qquad q_{xt} = 4jaq, \qquad r_{xt} = 4jar$$

Specializing

$$a = -\frac{j}{4}\cos u, \qquad b = c = -\frac{j}{4}\sin u, \qquad q = -r = -\frac{u_x}{2}$$

gives the sine Gordon equation

$$u_{xt} = \sin u$$

whereas

$$a = -\frac{j}{4}\cosh u, \qquad b = -c = -\frac{j}{4}\sinh u, \qquad q = r = \frac{u_x}{2}$$

gives the sinh Gordon equation

$$u_{xt} = \sinh u$$

There is still a three-step linear procedure to solve the original nonlinear equations, if one has found the correct specialization of Eqs. (E.17) and (E.18) (see Fig. E.1d):

1. Solve the linear problem Eq. (E.17) at $t = 0$ for the given initial values $q(x,0), r(x,0)$: direct scattering. This yields the spectral values $\lambda = \zeta$ and eigenfunctions $v(x,0)$, that is, the spectral data $S(t = 0)$.
2. Use the linear evolution Eq. (E.18) to get $v(x,t)$; λ is constant: $S(t)$.
3. Use $S(t)$ in order to calculate $F(x,t)$ or $q(x,t)$ and $r(x,t)$: inverse scattering (see Section E.4).

It is only in step 2 that "time" t plays a role. In steps 1 and 3, t is kept constant: $t = 0$ or $t > 0$, so that we can omit t in steps 1 and 3.

We shall now describe explicitly this procedure.

E.2   DIRECT SCATTERING; CALCULATION OF SCATTERING DATA

In what follows we shall assume that q and r vanish rapidly enough as $|x| \to \infty$. We define four solutions of the matrix differential equation (E.17) by their asymptotic behavior (for ζ real):

$$
\begin{aligned}
\Phi &= \mathbf{R}e^{j\zeta x}, & \mathbf{R} &\to \begin{bmatrix} 1 \\ 0 \end{bmatrix} & \text{as } x \to -\infty \\[2mm]
\overline{\Phi} &= -\overline{\mathbf{R}}e^{-j\zeta x}, & \overline{\mathbf{R}} &\to \begin{bmatrix} 0 \\ 1 \end{bmatrix} & \text{as } x \to -\infty \\[2mm]
\Psi &= \mathbf{L}e^{-j\zeta x}, & \mathbf{L} &\to \begin{bmatrix} 0 \\ 1 \end{bmatrix} & \text{as } x \to +\infty \\[2mm]
\overline{\Psi} &= \overline{\mathbf{L}}e^{+j\zeta x}, & \overline{\mathbf{L}} &\to \begin{bmatrix} 1 \\ 0 \end{bmatrix} & \text{as } x \to +\infty
\end{aligned}
\tag{E.28}
$$

We introduce the notation

$$
\mathbf{Q}(x) = \begin{bmatrix} 0 & q(x) \\ r(x) & 0 \end{bmatrix}, \qquad \mathbf{M}_+ = \begin{bmatrix} 0 & 0 \\ 0 & -2j \end{bmatrix}, \qquad \mathbf{M}_- = \begin{bmatrix} 2j & 0 \\ 0 & 0 \end{bmatrix}
$$

Then $\mathbf{R}, \overline{\mathbf{R}}, \mathbf{L}, \overline{\mathbf{L}}$ have to satisfy the following equations:

$$
\begin{aligned}
\partial_x \mathbf{R} &= (\zeta \mathbf{M}_+ + \mathbf{Q}) \cdot \mathbf{R}(x, \zeta) \\
\partial_x \overline{\mathbf{R}} &= (\zeta \mathbf{M}_- + \mathbf{Q}) \cdot \overline{\mathbf{R}}(x, \zeta) \\
\partial_x \mathbf{L} &= (\zeta \mathbf{M}_- + \mathbf{Q}) \cdot \mathbf{L}(x, \zeta) \\
\partial_x \overline{\mathbf{L}} &= (\zeta \mathbf{M}_+ + \mathbf{Q}) \cdot \overline{\mathbf{L}}(x, \zeta)
\end{aligned}
\tag{E.29}
$$

These matrix differential equations can be reformulated as integral equations of the Volterra type:

$$
\begin{aligned}
\mathbf{R}(x, \zeta) &= \begin{bmatrix} 1 \\ 0 \end{bmatrix} + \int_{-\infty}^{x} \mathbf{H}_1(x, y, \zeta) \mathbf{R}(y, \zeta)\, dy \\[2mm]
\overline{\mathbf{R}}(x, \zeta) &= \begin{bmatrix} 0 \\ 1 \end{bmatrix} + \int_{-\infty}^{x} \mathbf{H}_2(x, y, \zeta) \overline{\mathbf{R}}(y, \zeta)\, dy \\[2mm]
\mathbf{L}(x, \zeta) &= \begin{bmatrix} 0 \\ 1 \end{bmatrix} - \int_{x}^{\infty} \mathbf{H}_2(x, y, \zeta) \mathbf{L}(y, \zeta)\, dy \\[2mm]
\overline{\mathbf{L}}(x, \zeta) &= \begin{bmatrix} 1 \\ 0 \end{bmatrix} - \int_{x}^{\infty} \mathbf{H}_1(x, y, \zeta) \overline{\mathbf{L}}(y, \zeta)\, dy
\end{aligned}
\tag{E.30}
$$

with

$$\mathbf{H}_1(x,y,\zeta) = \begin{bmatrix} 0 & q(y) \\ r(y)\exp[-2j\zeta(x-y)] & 0 \end{bmatrix} = \exp[\zeta(x-y)\mathbf{M}_+]\cdot\mathbf{Q}(y)$$

$$\mathbf{H}_2(x,y,\zeta) = \begin{bmatrix} 0 & q(y)\exp[2j\zeta(x-y)] \\ r(y) & 0 \end{bmatrix} = \exp[\zeta(x-y)\mathbf{M}_-]\cdot\mathbf{Q}(y)$$

$$(E.31)$$

It is easy to show that for two arbitrary solutions $\mathbf{v}$ and $\mathbf{w}$ of Eq. (E.17),

$$\frac{\partial}{\partial x} W(\mathbf{v},\mathbf{w}) \equiv \frac{\partial}{\partial x}\det(\mathbf{v},\mathbf{w}) \equiv \frac{\partial}{\partial x}(v_1 w_2 - v_2 w_1) = 0$$

holds, so the Wronskian $W(\mathbf{v},\mathbf{w}) = $ const and can be computed at a convenient position $x = x_0$. $W \neq 0$ signifies linear independence, $W = 0$ linear dependence. Therefore, we calculate

$$\text{at } x \to -\infty: \quad W(\boldsymbol{\Phi},\overline{\boldsymbol{\Phi}}) = \begin{vmatrix} \exp(j\zeta x) & 0 \\ 0 & -\exp(-j\zeta x) \end{vmatrix} = -1$$

$$(E.32)$$

$$\text{at } x \to +\infty: \quad W(\boldsymbol{\Psi},\overline{\boldsymbol{\Psi}}) = \begin{vmatrix} 0 & \exp(j\zeta x) \\ \exp(-j\zeta x) & 0 \end{vmatrix} = -1$$

Therefore, $(\boldsymbol{\Phi},\overline{\boldsymbol{\Phi}})$ and $(\boldsymbol{\Psi},\overline{\boldsymbol{\Psi}})$ are, respectively, two sets of linear independent solutions, and we can write with four unknown complex functions $a, b, \bar{a}, \bar{b}$ (independent of x)

$$
\begin{array}{ccccc}
 & & x \to -\infty & & x \to +\infty \\[4pt]
\boldsymbol{\Phi} = a\,\overline{\boldsymbol{\Psi}} + b\,\boldsymbol{\Psi}, & \to & \begin{bmatrix} \exp(j\zeta x) \\ 0 \end{bmatrix}, & \to & \begin{bmatrix} a\exp(j\zeta x) \\ b\exp(-j\zeta x) \end{bmatrix} \\[18pt]
\overline{\boldsymbol{\Phi}} = \bar{b}\,\overline{\boldsymbol{\Psi}} - \bar{a}\,\boldsymbol{\Psi}, & \to & \begin{bmatrix} 0 \\ -\exp(-j\zeta x) \end{bmatrix}, & \to & \begin{bmatrix} \bar{b}\exp(j\zeta x) \\ -\bar{a}\exp(-j\zeta x) \end{bmatrix}
\end{array}
\quad (E.33)
$$

Then, with $W(\boldsymbol{\Phi},\overline{\boldsymbol{\Phi}}) = W(\boldsymbol{\Psi},\overline{\boldsymbol{\Psi}}) = -1$, we find immediately

$$a\bar{a} + b\bar{b} = 1 \tag{E.34}$$

Moreover,

$$a = W(\boldsymbol{\Phi},\boldsymbol{\Psi}), \quad b = -W(\boldsymbol{\Phi},\overline{\boldsymbol{\Psi}}), \quad \bar{a} = W(\overline{\boldsymbol{\Phi}},\overline{\boldsymbol{\Psi}}), \quad \bar{b} = W(\overline{\boldsymbol{\Phi}},\boldsymbol{\Psi}) \quad (E.35)$$

With Eq. (E.34) follows the inversion:

$$
\begin{array}{ccc}
 & x \to -\infty & x \to +\infty \\[6pt]
\mathbf{\Psi} = -a\overline{\mathbf{\Phi}} + \overline{b}\mathbf{\Phi}, \quad \to & \begin{bmatrix} \overline{b}\,\exp(j\zeta x) \\ a\,\exp(-j\zeta x) \end{bmatrix}, & \begin{bmatrix} 0 \\ \exp(-j\zeta x) \end{bmatrix} \\[14pt]
\overline{\mathbf{\Psi}} = b\overline{\mathbf{\Phi}} + \overline{a}\mathbf{\Phi}, \quad \to & \begin{bmatrix} \overline{a}\,\exp(j\zeta x) \\ -b\,\exp(-j\zeta x) \end{bmatrix}, & \begin{bmatrix} \exp(j\zeta x) \\ 0 \end{bmatrix}
\end{array}
\tag{E.36}
$$

These linear combinations may be interpreted as a scattering problem. If we take into account the time factor $\exp(j\omega t)$, we see that, for example, in the first Eq. (E.33) [*Note:* $\begin{bmatrix}1\\0\end{bmatrix}$] can be understood in quantum theory as representing positively charged particles, $\begin{bmatrix}0\\1\end{bmatrix}$ as negatively charged particles] $\begin{bmatrix}1\\0\end{bmatrix}e^{j\zeta x}$ represents positive particles receding to $-\infty$, and $b\begin{bmatrix}0\\1\end{bmatrix}e^{-j\zeta x}$ the flux of negative particles receding to $+\infty$, whereas $a\begin{bmatrix}1\\0\end{bmatrix}e^{j\zeta x}$ represents an incident flux of positive particles coming from $+\infty$. For an incoming unit flux of particles, we divide by a and define a transmission coefficient T and a reflection coefficient R as follows:

$$
T(\zeta) = \frac{1}{a(\zeta)}, \quad R(\zeta) = \frac{b(\zeta)}{a(\zeta)}
\tag{E.37}
$$

[*Note:* Do not mix the reflection coefficient R with the 2×1 matrix $\mathbf{R}(x,\zeta)$.]

Similarly, the second Eq. (E.33) also represents the scattering of particles with a transmission coefficient and a reflection coefficient:

$$
\overline{T}(\zeta) = \frac{1}{\overline{a}(\zeta)}, \quad \overline{R}(\zeta) = \frac{\overline{b}(\zeta)}{\overline{a}(\zeta)}
\tag{E.38}
$$

Now we shall extend our functions into the upper and lower half-plane of the complex ζ-plane (Im $\zeta \neq 0$). We state without proof that if

$$
\int_{-\infty}^{+\infty} |r(y)|\,dy < \infty, \qquad \int_{-\infty}^{+\infty} |q(y)|\,dy < \infty
$$

then $\mathbf{R}$, $\mathbf{L}$, and a are analytic in the lower half-plane, Im $\zeta < 0$, and $\overline{\mathbf{R}}$, $\overline{\mathbf{L}}$, $\overline{a}$ are analytic in the upper half-plane, Im $\zeta > 0$.

As long as q and r are not infinitesimally small, Eq. (E.17) can possess discrete eigenvalues (these represent bound states in quantum scattering theory). They are given by the poles of the transmission coefficients $T(\zeta)$ or the zeros of $a(\zeta)$: $a(\zeta = \zeta_j) = 0, j = 1, \ldots, N$ in the lower half-plane: Im $\zeta_j < 0$, and by the poles of $\overline{T}(\zeta)$ or the zeros $\overline{\zeta}_j$ of $\overline{a}(\zeta)$: $\overline{a}(\zeta = \overline{\zeta}_j) = 0, j = 1, \ldots, N$ for Im $\overline{\zeta}_j > 0$. Then we have instead of Eq. (E.33)

$$
\text{at } \zeta = \zeta_j: \quad \mathbf{\Phi} = C_j \mathbf{\Psi}, \qquad \text{at } \zeta = \overline{\zeta}_j: \quad \overline{\mathbf{\Phi}} = \overline{C}_j \overline{\mathbf{\Psi}}
\tag{E.39}
$$

If q and r decay sufficiently rapidly:

$$\int\limits_{-\infty}^{+\infty} |x|^n \left\{ \begin{matrix} |q(x)| \\ |r(x)| \end{matrix} \right\} dx < \infty \tag{E.40}$$

then we can also extend $b, \bar{b}$ into the complex ζ-plane and get $C_j = b(\zeta_j), \bar{C}_j = \bar{b}(\bar{\zeta}_j)$.

We still need the residues of $T(\zeta) = 1/a(\zeta)$ and $R(\zeta)$, respectively, at $\zeta = \zeta_j$ with $a(\zeta_j) = 0$. We find immediately (for simple poles)

$$\text{Res } T(\zeta_j) = \frac{1}{\partial a / \partial \zeta} \Big|_{\zeta = \zeta_j} =: \frac{1}{\dot{a}_j} \tag{E.41}$$

and the residues c_j of $R(\zeta)$ at $\zeta = \zeta_j$ are

$$c_j := \text{Res } R(\zeta_j) = \frac{b(\zeta_j)}{\dot{a}_j}$$

Now, the scattering data consist of the discrete eigenvalues ζ_j, the residues $c_j = b(\zeta_j)/\dot{a}_j$, and the reflection coefficient at $\xi = \text{Re } \zeta$:

$$S(t = 0) = \left\{ \{\zeta_j\}_{j=1}^N, \{c_j\}_{j=1}^N, R(\xi) = \frac{b(\xi)}{a(\xi)} \right\}_{t=0} \tag{E.42}$$

Similarly, the scattering data $\bar{S}(t = 0)$ are defined.

E.3   TIME EVOLUTION OF SCATTERING DATA

In this section, Eq. (E.18) must be solved. In our functions, we must add time t to the list of independent parameters, for example,

$$\Phi(x, \zeta) \to \Psi(x, t, \zeta), \qquad A(x, \zeta) \to A(x, t, \zeta), \qquad a(\zeta) \to a(t, \zeta)$$

In all our examples in Section E.1, we have

$$B(x, t, \zeta) \to 0, \qquad C(x, t, \zeta) \to 0, \qquad A(x, t, \zeta) \to A_\infty(\zeta)$$

as $|x| \to \infty$ [see Eqs. (E.22) and (E.25) with Eq. (E.40)]. These conditions are sufficient for our purposes; otherwise, the theory would be much more complicated. Whereas the expressions $\Phi, \bar{\Phi}, \Psi, \bar{\Psi}(x, t, \zeta)$ cannot be solutions of Eq. (E.18) (they would give no t-dependence; see the following discussion), the new functions

$$\mathbf{\Phi}^{(t)} = \mathbf{\Phi}\exp(A_\infty t), \qquad \overline{\mathbf{\Phi}}^{(t)} = \overline{\mathbf{\Phi}}\exp(-A_\infty t)$$
$$\mathbf{\Psi}^{(t)} = \mathbf{\Psi}\exp(-A_\infty t), \qquad \overline{\mathbf{\Psi}}^{(t)} = \overline{\mathbf{\Psi}}\exp(A_\infty t)$$

(E.43)

will do it. Inserting, for example, the first Eq. (E.43) into Eq. (E.18), we find

$$\frac{\partial}{\partial t}\mathbf{\Phi} = \begin{bmatrix} A - A_\infty & B \\ C & -A - A_\infty \end{bmatrix}\mathbf{\Phi}$$

Specializing to $x \to \infty$, we find [see Eq. (E.33)]

$$\begin{bmatrix} a_t\exp(j\zeta x) \\ b_t\exp(-j\zeta x) \end{bmatrix} = \begin{bmatrix} 0 & 0 \\ 0 & -2A_\infty \end{bmatrix}\begin{bmatrix} a\exp(j\zeta x) \\ b\exp(-j\zeta x) \end{bmatrix}$$

or

$$a_t = 0, \qquad b_t = -2A_\infty(\zeta)b$$

(E.44)

and

$$a(t,\zeta) = a(0,\zeta), \qquad b(t,\zeta) = b(0,\zeta)\cdot e^{-2A_\infty(\zeta)t}$$

(E.45)

So the eigenvalues ζ_j ($=$ zeros of a) are fixed in time. From Eq. (E.45) it follows that the residue $c_j(t)$ develops as

$$c_j(t) = \frac{b(t,\zeta_j)}{\dot{a}_j} = c_j(0)\cdot e^{-2A_\infty(\zeta_j)t} = c_j\cdot e^{-2A_\infty(\zeta_j)t}$$

(E.46)

Thus, the new scattering data are

$$S(t) = \left\{\{\zeta_j\}_{j=1}^N, \{c_j(t)\}_{j=1}^N, R(t,\xi) = \frac{b(t,\xi)}{a(0,\xi)} = \frac{b(t,\xi)}{a(\xi)}\right\}$$

Similarly, we find starting with $\mathbf{\Psi}^{(t)}$ from Eq. (E.43) the scattering data $\overline{S}(t)$:

$$\overline{a}(t,\zeta) = \overline{a}(0,\zeta), \qquad \overline{b}(t,\zeta) = \overline{b}(0,\zeta)\cdot e^{2A_\infty(\zeta)t}, \qquad \overline{c}(t,\overline{\zeta}_j) = \overline{c}(0,\overline{\zeta}_j)e^{2A_\infty(\overline{\zeta}_j)t}$$

(E.47)

E.4 THE INVERSE SCATTERING

We define two 2×1 matrices

$$\mathbf{K}(x,s) = \begin{bmatrix} K_1(x,s) \\ K_2(x,s) \end{bmatrix} \text{ and } \overline{\mathbf{K}}(x,s) = \begin{bmatrix} \overline{K}_1(x,s) \\ \overline{K}_2(x,s) \end{bmatrix}$$

with $\lim\limits_{s\to\infty} \mathbf{K}(x,s) = \lim\limits_{s\to\infty} \overline{\mathbf{K}}(x,s) = 0$ by the two equations

$$\Psi = \begin{pmatrix} 0 \\ 1 \end{pmatrix} e^{-j\zeta x} + \int_x^\infty \mathbf{K}(x,s)e^{-j\zeta s}\,ds$$

$$\overline{\Psi} = \begin{pmatrix} 1 \\ 0 \end{pmatrix} e^{j\zeta x} + \int_x^\infty \overline{\mathbf{K}}(x,s)e^{j\zeta s}\,ds \qquad \text{(E.48)}$$

We consider ζ on a contour C in the complex ζ-plane starting at $\zeta = -\infty - j0$, passing below all zeros of $a(\zeta)$, and ending at $\zeta = +\infty - j0$. The first of Eq. (E.33) can also be extended into the lower half-plane. After division with $a(\zeta)$, we get from it

$$T(\zeta)\Phi(x,\zeta) = \overline{\Psi}(x,\zeta) + R(\zeta)\Psi(x,\zeta) \qquad \text{(E.49)}$$

Substituting Eq. (E.48) into Eq. (E.49) gives

$$T(\zeta)\mathbf{R}e^{j\zeta x} = \begin{pmatrix} 1 \\ 0 \end{pmatrix} e^{j\zeta x} + \int_x^\infty \overline{\mathbf{K}}(x,s)e^{j\zeta s}\,ds + R(\zeta)\left[\begin{pmatrix} 0 \\ 1 \end{pmatrix} e^{-j\zeta x} + \int_x^\infty \mathbf{K}(x,s)e^{-j\zeta s}\,ds\right]$$

We operate on this equation with $1/2\pi \int_C d\zeta \cdot e^{-j\zeta y} \ldots$ for $y > x$ and obtain

$$0 = \overline{\mathbf{K}}(x,y) + \begin{pmatrix} 0 \\ 1 \end{pmatrix} F(x+y) + \int_x^\infty \mathbf{K}(x,s)F(s+y)\,ds \qquad \text{(E.50)}$$

where

$$F(x) = \frac{1}{2\pi}\int_C R(\zeta)e^{-j\zeta x}\,d\zeta \qquad \text{(E.51)}$$

while

$$\frac{1}{2\pi}\int_C T(\zeta)\mathbf{R}(x,\zeta)e^{j\zeta x}e^{-j\zeta y}\,d\zeta = \frac{1}{2\pi}\int_C \frac{\mathbf{R}}{a}e^{-j\zeta(y-x)}\,d\zeta = 0$$

Starting with the second of Eq. (E.33) yields similarly for a contour $\overline{C}$ above all zeros of $\overline{a}(\zeta)$ in the upper half-plane the integral equation $(y > x)$

$$0 = \mathbf{K}(x,y) - \begin{pmatrix} 1 \\ 0 \end{pmatrix} \overline{F}(x+y) - \int_x^\infty \overline{\mathbf{K}}(x,s)\overline{F}(s+y)\,ds \qquad \text{(E.52)}$$

and

$$\overline{F}(x) = \frac{1}{2\pi} \int\limits_{\overline{C}} \overline{R}(\zeta) e^{j\zeta x} d\zeta \tag{E.53}$$

If $a(\zeta)$ and $\overline{a}(\zeta)$ do not vanish on the real axis (Im $\zeta = 0, \zeta = \xi$), then we find by contour integration

$$F(x) = \frac{1}{2\pi} \int\limits_{-\infty}^{+\infty} R(\xi) e^{-j\xi x} d\xi + j \sum_{i=1}^{N} c(\zeta_i) \cdot e^{-j\zeta_i x}$$

$$\tag{E.54}$$

$$\overline{F}(x) = \frac{1}{2\pi} \int\limits_{-\infty}^{+\infty} \overline{R}(\xi) e^{+j\xi x} d\xi - j \sum_{i=1}^{\overline{N}} \overline{c}(\overline{\zeta}_i) \cdot e^{+j\overline{\zeta}_i x}$$

Knowledge of the scattering data $S(t), \overline{S}(t)$ allows us to therefore calculate F and $\overline{F}$ and from the integral equations (E.50) and (E.52)—in principle—**K** and **K̄**.

The two integral equations (E.50) and (E.52) can be written as a single matrix integral equation by defining

$$\mathcal{K} = \begin{bmatrix} \overline{K}_1 & K_1 \\ \overline{K}_2 & K_2 \end{bmatrix}, \qquad \mathcal{F} = \begin{bmatrix} 0 & -\overline{F} \\ F & 0 \end{bmatrix} \tag{E.55}$$

as follows:

$$\mathcal{K}(x,y) + \mathcal{F}(x,y) + \int\limits_{x}^{\infty} \mathcal{K}(x,s)\mathcal{F}(s+y)\, ds = 0 \tag{E.56}$$

These integral equations are named after Gelfand, Levitan and Marchenko.

In the last step we must now show how to recover the potentials $q(x)$ and $r(x)$ from **K** and **K̄**.

Substituting the first of Eq. (E.48) into Eq. (E.17), we find

$$\int\limits_{x}^{\infty} e^{-j\zeta s} [(\partial_x - \partial_s) K_1(x,s) - q(x) K_2(x,s)]\, ds = [q(x) + 2K_1(x,x)] e^{-j\zeta x}$$

$$\int\limits_{x}^{\infty} e^{-j\zeta s} [(\partial_x + \partial_s) K_2(x,s) - r(x) K_1(x,s)]\, ds = 0$$

These equations are fulfilled if

$$(\partial_x - \partial_s)K_1(x,s) - q(x)K_2(x,s) = 0$$
$$(\partial_x + \partial_s)K_2(x,s) - r(x)K_1(x,s) = 0$$
(E.57)

and

$$q(x) = -2K_1(x,x)$$
(E.58)

Similarly, we get from the second Eq. (E.48)

$$r(x) = -2\overline{K}_2(x,x)$$
(E.59)

These two last equations, (E.58) and (E.59), finally solve the nonlinear problem from which we started: the calculation of the two potential function $q(x)$ and $r(x)$.

E.5 SPECIALIZATION TO THE NONLINEAR SCHRÖDINGER EQUATION

We choose now in Eq. (E.17) $r = -q^*$ in order to find the NLS, Eq. (E.24),

$$jq_t(x,t) = q_{xx}(x,t) + 2|q(x,t)|^2 q(x,t)$$

If $\mathbf{v}$ is a solution of Eq. (E.17) at $\zeta_1 = \xi_1 + j\eta_1$, then it is easy to show that

$$\overline{\mathbf{v}} = \begin{bmatrix} v_2^* \\ -v_1^* \end{bmatrix}$$
(E.60)

is a solution at $\zeta_2 = \zeta_1^* = \xi_1 - j\eta_1$. So, if we start with the two solutions $\mathbf{\Phi}$ and $\mathbf{\Psi}$ of Eq. (E.28), we define $\overline{\mathbf{\Phi}}$ and $\overline{\mathbf{\Psi}}$ as

$$\overline{\mathbf{\Phi}}(x,\zeta) = \begin{bmatrix} \Phi_2^*(x,\zeta^*) \\ -\Phi_1^*(x,\zeta^*) \end{bmatrix}, \qquad \overline{\mathbf{\Psi}}(x,\zeta) = \begin{bmatrix} \Psi_2^*(x,\zeta^*) \\ -\Psi_1^*(x,\zeta^*) \end{bmatrix}$$
(E.61)

which implies

$$\overline{a}(\zeta) = a^*(\zeta^*), \qquad \overline{b}(\zeta) = b^*(\zeta^*)$$
(E.62)

and consequently,

$$N = \overline{N}, \qquad \overline{\zeta}_j = \zeta_j^*, \qquad \overline{c}_j = c_j^*$$

From Eq. (E.54), we now find

$$\overline{F}(x) = F^*(x)$$
(E.63)

and from a comparison of Eqs. (E.50) and (E.52), we get

$$\overline{\mathbf{K}}(x,y) = \begin{bmatrix} K_2^*(x,y) \\ -K_1^*(x,y) \end{bmatrix} \tag{E.64}$$

The Marchenko integral equations (E.50) now transform into

$$0 = + K_2^*(x,y) + \int_x^\infty K_1(x,s)F(s+y)\,ds$$

$$\tag{E.65}$$

$$0 = - K_1^*(x,y) + \int_x^\infty K_2(x,s)F(s+y)\,ds + F(x+y)$$

We start with a solution of the differential equation (E.17) with $r = -q^*$ and ζ real and construct explicitly the scattering data. Eliminating v_2 in (E.17) yields for the upper component v_1 of **v**

$$v_{1xx} + (\zeta^2 + |q|^2)v_1 + (j\zeta v_1 - v_{1x}) \cdot \frac{q_x}{q} = 0 \tag{E.66}$$

In the first step, we are interested in a solution for $t = 0$, where we now take in particular as the initial condition [see, e.g., Eq. (10.16)]

$$q(x, t = 0) = q_0 \cdot \operatorname{sech} x, \qquad q_0 > 0 \tag{E.67}$$

and the start value of the potential $q(x,0)$ is then real. With the new independent variable

$$s = \frac{1 - \tanh x}{2} \tag{E.68}$$

we find from Eq. (E.66)

$$s(1-s)v_{1ss} + \frac{1-2s}{2}v_{1s} + \left[q_0^2 + \frac{\zeta^2 - j\zeta(1-2s)}{4s(1-s)}\right]v_1 = 0 \tag{E.69}$$

Similarly, the differential equation for the lower component v_2 of **v** is found by replacing ζ by $-\zeta$ in Eq. (E.69). One solution of both ODEs can be expressed as follows with hypergeometric functions $F(a, b; c; s)$ (Abramowitz, 1968):

$$v_1^{(1)} = s^{-j\zeta/2}(1-s)^{j\zeta/2}F\left(-q_0, q_0; \frac{1}{2} - j\zeta; s\right)$$

$$\tag{E.70}$$

$$v_2^{(1)} = s^{j\zeta/2}(1-s)^{-j\zeta/2}F\left(-q_0, q_0; \frac{1}{2} + j\zeta; s\right)$$

The second linear independent solution of the ODEs for v_1 and v_2, respectively, is given by

$$v_1^{(2)} = s^{-j\zeta/2}(1-s)^{j\zeta/2} \cdot s^{1/2+j\zeta} F\left(\frac{1}{2}+j\zeta-q_0, \frac{1}{2}+j\zeta+q_0; \frac{3}{2}+j\zeta; s\right)$$

$$= s^{1/2+j\zeta/2}(1-s)^{j\zeta/2} F\left(\frac{1}{2}+j\zeta-q_0, \frac{1}{2}+j\zeta+q_0; \frac{3}{2}+j\zeta; s\right) \qquad \text{(E.71)}$$

$$v_2^{(2)} = s^{1/2-j\zeta/2}(1-s)^{-j\zeta/2} F\left(\frac{1}{2}-j\zeta-q_0, \frac{1}{2}-j\zeta+q_0; \frac{3}{2}-j\zeta; s\right)$$

Using the following two properties of the hypergeometric functions:

$$F(\alpha,\beta;\gamma;s=0)=1, \qquad F(\alpha,\beta;\gamma;s=1)=\frac{\Gamma(\gamma)\Gamma(\gamma-\alpha-\beta)}{\Gamma(\gamma-\alpha)\Gamma(\gamma-\beta)}$$

we now construct the solution $\boldsymbol{\Psi}=(\Psi_1,\Psi_2)^T$, Eq. (E.36), from the functions $v_i^{(j)}$. From the asymptotic behavior as $x\to\infty$ $(s\to 0)$, we have to choose

$$\Psi_1=\rho v_1^{(2)}, \qquad \Psi_2=v_2^{(1)} \qquad \text{(E.72)}$$

where ρ is a constant factor, and from $x\to-\infty$ $(s\to 1)$,

$$a=\frac{\Gamma^2(1/2+j\zeta)}{\Gamma(1/2+j\zeta+q_0)\Gamma(1/2+j\zeta-q_0)}, \qquad |a|^2=1-\frac{\sin^2\pi q_0}{\cosh^2\pi\zeta} \qquad \text{(E.73)}$$

If we choose $\rho=q_0/(\zeta-j/2)$, or

$$b^*=\rho\cdot\frac{\Gamma(3/2+j\zeta)\Gamma(1/2-j\zeta)}{\Gamma(1+q_0)\Gamma(1-q_0)}=j\frac{\Gamma(1/2-j\zeta)\Gamma(1/2+j\zeta)}{\Gamma(q_0)\Gamma(1-q_0)}=j\frac{\sin\pi q_0}{\cosh\pi\zeta} \qquad \text{(E.74)}$$

then Eq. (E.34), $|a|^2+|b|^2=1$ is satisfied. Analytic continuation into the lower ζ half-plane leads to the zeros of $a(\zeta)$ or poles of $\Gamma(1/2+j\zeta-q_0)$:

$$\zeta=\zeta_r=0+j\eta_r=j\left(r-\frac{1}{2}-q_0\right), \qquad r<\frac{1}{2}+q_0, \qquad r=1,2,3,\ldots \qquad \text{(E.75)}$$

all lying on the negative imaginary axis.

We now take $q_0=N$ $(=1,2,3,\ldots)$. Then the condition $r<N+1/2$ is fulfilled exactly for the N values $r=1,\ldots,N$: there are N bound states representing an N-soliton. In this case, we see from Eq. (E.74) that for $\zeta=\xi$ (real), there are no reflections:

$$\frac{1}{\Gamma(1-N)}=0, \qquad b(\xi)=0, \qquad R(\xi)=\frac{b(\xi)}{a(\xi)}=0 \qquad \text{(E.76)}$$

while at the complex zeros ζ_r,

$$b(\zeta = \zeta_r) = -j\frac{\sin \pi N}{\cosh j\pi(r - 1/2 - N)} = j(-1)^r \qquad (\text{E.77})$$

Moreover,

$$\sum_{r=1}^{N} |2\eta_r| = N^2 = q_0^2 \qquad (\text{E.78})$$

This power of the N-soliton solution is thus N^2 times the power of the single soliton with $N = 1$; see Eq. (E.67).

We have here a reflectionless potential: $R(t = 0, \xi) = 0$ (no continuous spectrum). Therefore, we take for $F(x)$, Eq. (E.54), in the N-soliton case

$$F(x) = j\sum_{i=1}^{N} c_i(t)e^{-j\zeta_i x}, \qquad \zeta_i = \xi_i + j\eta_i, \qquad \xi_i = 0, \quad \eta_i < 0, \qquad a(\zeta_i) = 0$$

$$(\text{E.79})$$

If we introduce two $(N \times 1)$ matrices $\mathbf{a}, \mathbf{b}$ [$()^T$ means transposition]

$$\mathbf{a}(z) = (e^{-j\zeta_1 z}, \ldots, e^{-j\zeta_N z})^T$$
$$\mathbf{b}(z) = [jc_1(t)e^{-j\zeta_1 z}, \ldots, jc_N(t)e^{-j\zeta_N z}]^T \qquad (\text{E.80})$$

which are known from the scattering data, we find

$$F(x + y) = \mathbf{b}^T(x)\mathbf{a}(y) \qquad (\text{E.81})$$

The solution of Eqs. (E.65) is easy only if $K_i(x, y)$ is separable. Therefore, we set

$$K_i(x, y) = \mathbf{k}_i^T(x)\mathbf{a}^*(y)$$

and $\mathbf{k}_i$ is still an unknown column matrix. We finally introduce the $(N \times N)$ matrix

$$\mathbf{M}(x) = \int_x^{\infty} \mathbf{a}^*(s)\mathbf{b}^T(s)\, ds \qquad (\text{E.82})$$

Then, we obtain from Eq. (E.65)

$$\mathbf{k}_1^\dagger = \mathbf{b}^T(x)(\mathbf{I} + \mathbf{M}^*\mathbf{M})^{-1}$$
$$\mathbf{k}_2^\dagger = -\mathbf{k}_1^T(x)\mathbf{M}(x) \qquad (\text{E.83})$$

[()† means transposition and complex conjugation, $\mathbf{I}$ is the $(N \times N)$ unit matrix.] Thus,

$$q(x) = -2K_1(x, x) = -2\mathbf{k}_1^T(x)\mathbf{a}^*(x) = -2\mathbf{b}^\dagger(x)(\mathbf{I} + \mathbf{MM}^*)^{-1}\mathbf{a}^*(x) \quad (E.84)$$

This is a closed-form solution of the NLS for an N-soliton.

For a single soliton with

$$N = 1, \qquad \zeta_1 = j\eta_1 = -\frac{j}{2}, \qquad \xi_1 = 0, \qquad a = \frac{\zeta - \zeta_1}{\zeta - \zeta_1^*}, \qquad \dot{a}_{|\zeta_1} = j$$

we have as scattering data

$$S(t = 0) = \left\{ \zeta_1 = -\frac{j}{2}, c_1 = -1, R(\xi) = 0 \right\} \quad (E.85)$$

From Eq. (E.22), we see that (with $a_2 = 2j$)

$$\lim_{|x| \to \infty} A(x, t, \zeta_1) = A_\infty(\zeta_1) = 2j\zeta_1^2 = -\frac{j}{2}, \qquad A_\infty(\xi) = 0 \quad (E.86)$$

and, according to Eq. (E.46),

$$c_1(t) = c(t, \zeta_1) = c_1 e^{-2A_\infty(\zeta_1)t} = -e^{jt} \quad (E.87)$$

This results in

$$S(t) = \left\{ \zeta_1 = -\frac{j}{2}, c_1(t) = -e^{jt}, R(t, \xi) = 0 \right\} \quad (E.88)$$

From Eqs. (E.69) and (E.82), we get

$$\mathbf{M} = \int_x^\infty e^{+j\zeta_1^* s} j c_1(t) e^{-j\zeta_1 s} ds = -j c_1(t) \frac{\exp(2\eta_1 x)}{2\eta_1} = -j e^{jt} e^{-x} \quad (E.89)$$

and from Eq. (E.84),

$$q(x, t) = e^{-jt} \text{sech } x \quad (E.90)$$

where we have omitted a constant phase factor since with q also $q \exp(j\psi)$, where $\psi = $ const is a solution of the NLS (E.24).

Now the ODE, Eq. (E.24), for the NLS is invariant with respect to translations in $x: x \to x - x_0$. Therefore,

$$q(x, t) = e^{-jt} \text{sech}(x - x_0) \quad (E.91)$$

is a solution too. But, Eq. (E.24) is also invariant with respect to a pure Galilean transformation:

$$x \rightarrow x' = -vt$$
$$t \rightarrow t' = t \qquad\qquad (E.92)$$

and the solution $q(x, t)$ acquires an additional phase factor

$$q(x, t) \rightarrow q'(x', t') = q(x, t) \exp\left[-j\frac{v}{2}\left(x - \frac{vt}{2}\right)\right] \qquad (E.93)$$

We set $v = 4\chi$ and get

$$q(x, t) = e^{4jt\chi^2 - 2jxx} e^{-jt} \text{sech}[x - (x_0 - 4\chi t)] \qquad (E.94)$$

This wave is moving with a phase velocity $v = (4\chi^2 - 1)/(2\chi)$. While the form of the solution remains stable, the peak value is shifted according to $x_0 \rightarrow x_0(t) = x_0 - 4\chi t$, with a group velocity of 4χ.

For $N = 2$ with $q(x, 0) = 2 \text{ sech } x$, we get

$$q(x, t) = 4e^{-jt} \frac{\cosh 3x + 3\exp(-8jt)\cosh x}{\cosh 4x + 4\cosh 2x + 3\cos 8t} \qquad (E.95)$$

Here, $|q(x, t)|^2$ is periodic in t with a period of $t_0 = \pi/4$. Figure E.2 shows $|q(x, t)|$ for the $N = 2$-soliton as a function of x for $t = 0, \pi/16, \pi/8, 3\pi/16, t_0$. One can

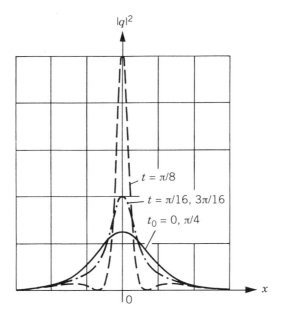

FIGURE E.2 Temporal behavior for the $N = 2$-soliton during one soliton period between $t = 0$ and $t = \pi/4$.

see how the pulse splits after half a period and recovers its initial shape after a full period $t = t_0$.

If $q_0 = N + \alpha$, $|\alpha| < 1/2$, the solution consists of N bound solitons plus a radiative part that decays as $t^{-1/2}$.

PROBLEMS

Section E.1

E.1 Show that with the Lax pair

$$L = -j\begin{bmatrix} 1+p & 0 \\ 0 & 1-p \end{bmatrix}\frac{\partial}{\partial x} + \begin{bmatrix} 0 & u^* \\ u & 0 \end{bmatrix}$$

$$M = -jp\begin{bmatrix} 1 & 0 \\ 0 & 1 \end{bmatrix}\frac{\partial^2}{\partial x^2} + \begin{bmatrix} \dfrac{j|u|^2}{1+p} & u_x^* \\ -u_x & -\dfrac{j|u|^2}{1-p} \end{bmatrix}$$

and $G = 2/(1-p^2)$, one finds the NLS in the form

$$j\frac{\partial u}{\partial t} - \frac{\partial^2 u}{\partial x^2} - G|u|^2 u = 0$$

Section E.2

E.2 Derive the matrix differential equations (E.29).

E.3 Derive the matrix integral equations (E.30).

Section E.5

E.4 Show the Galilean invariance of Eq. (E.24), that is, if we use the transformations $x \to x'$, $t \to t'$, $q(x,t) \to q'(x',t')$ as in Eqs. (E.92) and (E.93), then we find the form-invariant equation.

E.5 Derive the $N = 2$-soliton solution Eq. (E.95).

E.6 Show that the solution, Eq. (E.95), reduces for $t = 0$ to $q(x,0) = 2\,\text{sech}\,x$.

REFERENCES

M. Abramowitz and I. A. Stegun (eds.). *Handbook of Mathematical Functions.* Dover, New York, 1968.

BIBLIOGRAPHY

M. J. Ablowitz, D. J. Kaup, A. C. Newell and H. Segur. The inverse scattering transform—Fourier analysis for nonlinear problems. *Studies Appl. Math.* 53 (1974): 249–315.

M. J. Ablowitz and H. Segur. *Solitons and the Inverse Scattering Transform.* SIAM, Philadelphia, PA, 1981.

A. P. Fordy (ed.). *Soliton Theory: a Survey of Results.* Manchester University Press, Manchester, UK, 1990.

A. Hasegawa. *Optical Solitons in Fibers.* Springer-Verlag, Berlin, 1989.

G. L. Lamb, Jr. *Elements of Soliton Theory.* Wiley, New York, 1980.

J. Satsuma and N. Yajima. Initial value problem of one-dimensional self-modulation of nonlinear waves in dispersive media. *Progr. Theor. Phys.* 55 (1974): 284–306 (suppl.).

V. E. Zakharov and A. B. Shabat. Exact theory of two-dimensional self-focusing and one-dimensional self-modulation of waves in nonlinear media. *Sov. Phys. JETP* 34 (1972): 62–69.

V. E. Zakharov and A. B. Shabat. Interaction between solitons in a stable medium. *Sov. Phys.* JETP 37 (1973): 823–828.

Index

The page numbers in **boldface** are references to the most important positions.

211